K. Berg V. Bulyzhenkov Y. Christen
P. Corvol (Eds.)

Genetic Approaches to Coronary Heart Disease and Hypertension

With 45 Figures and 29 Tables

Springer-Verlag
Berlin Heidelberg New York
London Paris Tokyo
Hong Kong Barcelona
Budapest

Berg, Kare, M. D., Ph. D.
University of Oslo
Institute of Medical Genetics
UllevalHospital, Oslo
Norway

Bulyzhenkov, Victor, Dr.
World Health Organization
Avenue Appia
1211 Geneva 27
Switzerland

Christen, Yves, Dr.
Fondation IPSEN pour la Recherche
Thérapeutique
30, rue Cambronne
F-75737 Paris CEDEX
France

Corvol, Pierre, M. D.
INSERM Unit 36
Collège de France
3 rue d'Ulm
F-75005 Paris

ISBN-13:978-3-642-76893-4 e-ISBN-13:978-3-642-76891-0
DOI: 10.1007/978-3-642-76891-0

Softcover reprint of the hardcover 1st edition 1991

2127/3140/543210 – Printed on acid-free paper

ALLE ZEIT WACH 1842

Preface

Cardiovascular diseases are the principal cause of mortality in many developed countries and are increasing in importance in developing countries where communicable diseases are being brought under control. It is estimated that coronary heart disease (CHD) and cerebrovascular diseases are responsible for between 40–50% of all deaths. High blood pressure is common in all industrialized societies and contributes importantly to CHD.

It is well known that genetic factors play an important role in the etiology and pathogeneses of cardiovascular diseases and contribute to an individual's susceptibility or resistance to the disease. The application of modern genetic techniques has resulted in the identification of polymorphic markers that are involved in the normal regulation and function of the cardiovascular system. Many of these may contribute to the individual differences in the risk for CHD and hypertension. It is now known that several risk factors or protective factors with respect to cardiovascular diseases are strongly influenced by genes, and it is now possible to identify individual genes contributing to cardiovascular risk. A large bank of genetic knowledge relevant to CHD and hypertension has become available during the past years. As a result, it is now believed that CHD, for example, could be a good model for a feasibility study to determine a way by which preventive measures can be applied.

In attempting to realize the genetic risk approach, methods should be available to identify individuals at risk before manifestation of the disease takes place. Thus, the early identification of persons in childhood and young adult life who are at high genetic risk constitutes a very powerful health care strategy for the prevention of CHD and hypertension. Clearly, if the high risk individuals could be identified unambiguously, a cost-effective prevention strategy would immediately become available. Progress in the DNA field through the development of DNA markers will probably lead to increased possibilities to identify persons with a genetic susceptibility to the disease, and may add a new "individual or family-oriented" element to preventive medicine.

Thus, in order to review and discuss fresh genetic data for the prevention and control of CHD and hypertension, as well as the place of the genetic approach within the integrated ones for community control of cardiovascular diseases, a joint meeting was initiated by the World Health Organization and the IPSEN Foundation in Moscow on December 3-4, 1990. The papers presented during the meeting were considered for further revision of the problems, and are presented here to the scientific community, in the hope that they will stimulate the exchange of information and further progress of knowledge about the genetics of cardiovascular diseases.

Although the papers included in this volume have been produced in the framework of WHO/IPSEN Foundation's review of knowledge about the genetics of mental disorders, they express the views of the individual authors rather than a consensus of the participants at the meeting. Also, the views do not necessarily represent the decision for stated policy of the World Health Organization or the IPSEN Foundation.

The editors wish to express their appreciation to Mrs J. Mervaillie for the organization of the meeting and Mrs M.-L. Gage for her editorial assistance.

K. Berg · V. Bulyzhenkov · Y. Christen · P. Corvol

Contents

Genetics of Hypertension – An Overview
A.G. Motulsky and *W. Burke* 1

A Population Perspective for Genetics Research and Applications to Control Cardiovascular Disease in Utah
R.R. Williams, S.C. Hunt, S.J. Hasstedt, P.N. Hopkins, L.L. Wu, M.C. Schumacher, T.D. Berry, B.M. Stults, G.K. Barlow, R.P. Lifton, and *J.-M. Lalouel* 8

Cell Membrane Alteration in Primary Hypertension: – An Evidence of Its Genomic Source
Y.V. Postnov 20

Molecular Genetics and Familial Arterial Hypertension
F. Soubrier, X. Jeunemaître, B. Rigat, F. Cambien, and *P. Corvol* . . 27

Gene Mapping in Experimental Hypertension
T.W. Kurtz, L. Simonet, V. Křen, and *M. Pravanec* 38

Transgenic Mice and the Development of Animal Models and Resources for Hypertension Research
C.D. Sigmund, C.A. Jones, J.R. Fabian, C. Wu, C.M. Kane, M.K. Ellsworth, F.D. Pacholec, and *K.W. Gross* 60

Tissue Distribution of Ren-2 Expression in Transgenic Rats
J. Peters, M. Bader, D. Ganten, and *J. Mullins* 74

Prediction of Hypertension and Stroke for Controlling Cardiovascular Diseases in Genetic Rat Models
Y. Yamori, Y. Nara, T. Nabika, K. Ikeda, and *K. Matsumoto* 81

An Overview of the Genetics of Coronary Heart Disease and Its Risk Factors
K. Berg 98

Genetic Risk Factors for Coronary Heart Disease
in the Japanese Population
H. Hamaguchi, S. Kikuchi, H. Yanagi, T. Matsunaga, Y. Yamanouchi, A. Nakagawa, Y. Watanabe, T. Tanaka, Y. Shimakura, N. Imoto, T. Arinami, and *K. Yamakawa-Kobayashi* 110

Genetic Epidemiology of Ischemic Heart Disease
and Arterial Hypertension
V.A. Koshechkin, C.B. Hudaiberdyev, V.A. Kabakova, S.V. Panin, T.A. Rozhkova, and *E.Y. Solovjova* 127

The Role of Lp(a) in Atherosclerosis
R. Lawn 143

Liver-Directed Gene Therapy in the Treatment
of Familial Hypercholesterolemia
J.M. Wilson and *M. Grossman* 152

Subject Index 157

Contributors

Arinami, T.
Department of Human Genetics, Institute of Basic Medical Sciences, University of Tsukuba, Tsukuba 305, Japan

Bader, M.
Institut für Bluthochdruckforschung und Institut für Pharmakologie, Universität Heidelberg, Im Neunheimer Feld 366, D–6900 Heidelberg

Barlow, G.K.
Department of Medicine, University of Utah Medical School, Salt Lake City, UT 84108, USA

Berg, K.
University of Oslo, Institute of Medical Gentics, Ulleval Hospital, Oslo, Norway

Berry, T.D.
Department of Medicine, University of Utah Medical School, Salt Lake City, UT 84108, USA

Burke, W.
Department of Medicine and Genetics, University of Washington, Seattle, WA 98195, USA

Cambien, F.
INSERM U 258, Hôpital Broussais, 96 rue Didot, F–75014 Paris

Corvol, P.
INSERM Unit 36, Collège de France, 3 rue d'Ulm, F–75005 Paris

Ellsworth, M.K.
Department of Molecular and Cellular Biology, Roswell Park Cancer Institute, Elm and Carlton Streets, Buffalo, NY

Fabian, J.R.
Department of Molecular and Cellular Biology, Roswell Park Cancer Institute, Elm and Carlton Streets, Buffalo, NY 14263, USA

Ganten, D.
Institut für Bluthochdruckforschung und Institut für Pharmakologie, Universität Heidelberg, Im Neunheimer Feld 366, D-6900 Heidelberg

Gross, K.W.
Department of Molecular and Cellular Biology, Roswell Park Cancer Institute, Elm and Carlton Streets, Buffalo, NY 14263, USA

Grossman, M.
Departments of Internal Medicine and Biological Chemistry, University of Michigan, Ann Arbor, MI 48109, USA

Hamaguchi, H.
Department of Human Gentics, Institute of Basic Medical Sciences, University of Tsukuba, Tsukuba 305, Japan

Hasstedt, S.J.
Department of Human Genetics, University of Utah Medical School, Salt Lakc City, UT 84108, USA

Hopkins, P.N.
Department of Medicine, University of Utah Medical School, Salt Lake City, UT 84108, USA

Hudaiberdyev, C.B.
USSR Academy of Medical Sciences, Cardiology Research Center, 121552 Moscow, USSR

Hunt, S.C.
Department of Medicine, University of Utah Medical School, Salt Lake City, UT 84108, USA

Ikeda, K.
Department of Pathology, Shimane Medical University, Izumo 693, Japan

Imoto, N.
Department of Human Genetics, Institute of Basic Medical Sciences, University of Tsukuba, Tsukuba 305, Japan

Jeunemaître, X.
INSERM Unit 36, Collège de France, 3 rue d'Ulm, F-75005 Paris

Jones, C.A.
Department of Molecular and Cellular Biology, Roswell Park Cancer Institute, Elm and Carlton Streets, Buffalo, NY 14263, USA

Kabakova, V.A.
USSR Academy of Medical Sciences, Cardiology Research Center, 121552 Moscow, USSR

Kane, C.M.
Department of Molecular and Cellular Biology, Roswell Park Cancer Institute, Elm and Carlton Streets, Buffalo, NY 14263, USA

Kikuchi, S.
Department of Human Genetics, Institute of Basic Medical Sciences, University of Tsukuba, Tsukuba 305, Japan

Koshechkin, V.
USSR Academy of Medical Sciences, Cardiology Research Center, 121552 Moscow, USSR

Kren, V.
Department of Biology, Faculty of General Medicine, Charles University, Prague, Czechoslovakia

Kurtz, T.W.
Department of Laboratory Medicine, University of California, Box 0134, San Francisco, CA 94143 – 0134, USA

Lalouel, J.-M.
Department of Medicine, University of Utah Medical School, Salt Lake City, UT 84108, USA

Lawn, R.
Stanford University School of Medicine, 300 Pasteur Drive, CVRC 267, Stanford, CA 94305–5246, USA

Lifton, R.P.
Brigham and Women's Hospital and Harvard Medical, Boston, MA 02115, USA

Matsumoto, K.
Institute for Animal Experimentation, University of Tokushima School of Medicine, Tokushima 770, Japan

Matsunaga, T.
Department of Human Genetics, Institute of Basic Medical Sciences, University of Tsukuba, Tsukuba 305, Japan

Motulsky, A.G.
Department of Medicine and Genetics, University of Washington, Seattle, WA 98195, USA

Mullins, J.
AFRC Centre for Genome Research, University of Edinburgh, King's Building, West Mains Road, Edinburgh EH9 3JQ, UK

Nabika, T.
Department of Pathology, Shimane Medical University, Izumo 693, Japan

Nakagawa, A.
Department of Human Genetics, Institute of Basic Medical Sciences, University of Tsukuba, Tsukuba 305, Japan

Nara, Y.
Department of Pathology, Shimane Medical University, Izumo 693, Japan

Pacholec, F.D.
Department of Molecular and Cellular Biology, Roswell Park Cancer Instittute, Elm and Carlton Streets, Buffalo, NY 14263, USA

Panin, S.V.
USSR Academy of Medical Sciences, Cardiology Research Center, 121552 Moscow, USSR

Peters, J.
Institut für Bluthochdruckforschung und Institut für Pharmakologie, Universität Heidelberg, Im Neunheimer Feld 366, D–6900 Heidelberg

Postnov, Y.V.
National Cardiology Research Centre, Moscow 121552, USSR

Pravenec, M.
Institute of Physiology, Czechoslovak Academy of Sciences, Prague, Czechoslovakia

Rigat, B.
INSERM Unit 36, Collège de France, 3 rue d'Ulm, F–75005 Paris

Rozhkova, T.A.
USSR Academy of Medical Sciences, Cardiology Research Center, 121552 Moscow, USSR

Schumacher, M.C.
Department of Medicine, University of Utah Medical School, Salt Lake City, UT 84108, USA

Shimakura, Y.
Department of Human Genetics, Institute of Basic Medical Sciences, University of Tsukuba, Tsukuba 305, Japan

Sigmund C.D.
Cardioviscular Divisior department of Internal Medicine University of Iowa College of Medicine Iowa City, Iowa 52242, USA

Simonet, L.
Department of Laboratory Medicine, University of California, Box 0134, San Francisco, CA 94143 – 0134, USA

Solovjova, E.Y.
USSR Academy of Medical Sciences, Cardiology Research Center, 121552 Moscow, USSR

Soubrier, F.
INSERM Unit 36, Collège de France, 3 rue d'Ulm, F–75005 Paris

Stults, B.M.
Department of Medicine, University of Utah Medical School, Salt Lake City, UT 84108, USA

Tanaka, T.
Department of Human Genetics, Institute of Basic Medical Sciences, University of Tsukuba, Tsukuba 305, Japan

Watanabe, Y.
Department of Human Genetics, Institute of Basic Medical Sciences, University of Tsukuba, Tsukuba 305, Japan

Williams, R.R.
University of Utah Cardiovascular Genetics Research Clinic, 410 Chipeta Way, Salt Lake City, UT 84108, USA

Wilson, J.M.
Departments of Internal Medicine and Biological Chemistry, University of Michigan, Ann Arbor, MI 48109, USA

Wu, C.
Department of Molecular and Cellular Biology, Roswell Park Cancer Institute, Elm and Carlton Streets, Buffalo, NY 14263, USA

Wu, L.L.
Department of Medicine, University of Utah Medical School, Salt Lake City, UT 84108, USA

Yamakawa-Kobayashi, K.
Department of Human Genetics, Institute of Basic Medical Sciences, University of Tsukuba, Tsukuba 305, Japan

Yamanouchi, Y.
Department of Human Genetics, Institute of Basic Medical Sciences, University of Tsukuba, Tsukuba 305, Japan

Yamori, Y.
Department of Pathology, Shimane Medical University, Izumo 693, Japan

Yanagi, H.
Department of Human Gentics, Institute of Basic Medical Sciences, University of Tsukuba 305, Japan

Genetics of Hypertension – An Overview

A.G. Motulsky and *W. Burke*

Summary

Hypertension represents the upper range of the continuous distribution of blood pressure seen in human populations. High blood pressure is a multifactorial trait requiring both genetic and environmental factors for its manifestation. Genetic heterogeneity is likely. Certain underdeveloped populations have no hypertension but some individuals from such groups have the genetic make-up to develop high blood pressure when exposed to Western-style-environments. The specific genes involved in human hypertension are largely unknown. Genetic studies of blood pressure require attention to intermediate phenotypes with effects that are close to the mechanisms involved in blood pressure regulation. Ion exchange across cell membranes in kidney and/or smooth muscle cells may be involved, as shown by studies of red cell sodium-lithium countertransport (CT). A unique single gene appears to be responsible for elevated CT and may contribute as much as one-fifth of the attributable risk for systolic hypertension. Genetic animal models for hypertension may not reflect the genes operative in human hypertension.

Hypertension is a common trait with familial aggregation (see Ward 1990; Burke and Motulsky 1991, for extensive literature citations). Its genetic etiology appears to be heterogeneous. Rarely, a monogenic condition such as polycystic kidneys may be the cause of hypertension. The cut-off point of 140/90 mm to 160/90 mmHg to define high blood pressure is somewhat arbitrary and is selected to designate individuals at higher risk for renal and vascular complications. The distribution of blood pressure in populations approximates a Gaussian distribution. The shape of this curve explains why the majority of the 15 %–20 % of persons designated as hypertensives will have only mild or moderate hypertension. Hypertension by itself is not a disease but represents a risk factor for stroke, congestive heart failure, coronary heart disease, and renal failure. Not every person – even with severe hypertension – develops these medical complications. We do not know what factors determine which of these complications will develop in a given patient. Genetic factors might explain the predilection for a certain complication such as strokes among some families.

Using the diagnosis of hypertension by itself as a clinical entity is hazardous for genetic studies because of the heterogeneous etiology of high blood pressure.

K. Berg, V. Bulyzhenkov, Y. Christen, P. Corvol (Eds.)
Genetic Approaches to Coronary Heart Disease and Hypertension

Findings that apply to one family may not apply to others and it is difficult to draw meaningful conclusions if the underlying biologic determinants vary between families.

A genetic approach to hypertension requires attention to the pathophysiologic and biochemical mechanisms that lead from the genotype to the phenotype of high blood pressure. Hypertension itself – as determined by a blood pressure reading – is far removed from gene action, and accurate measurement of blood pressure is dependent on the use of careful techniques. The study of phenotypic effects "closer" to the mechanisms involved in blood pressure regulation is more likely to yield insights into the role of genes that determine high blood pressure.

Familial Aggregation Studies

Many studies have been done to assess the correlation of systolic and diastolic blood pressure in various classes of biologic and nonbiologic relatives (see Ward 1990; Burke and Motulsky 1991). Identical twins who share all their genes resemble each other more closely in blood pressure than nonidentical twins who share 50 % of their genes. First-degree relatives such as sibs or parents and their children have more similar blood pressure than sets of adopted children and their nonbiologic parents or adopted sibs living in the same family. Biologic sibs after many years of separation still show correlations for blood pressure somewhat similar to those of sibs living together. Conversely, spouses living together for many years show insignificant correlations for blood pressure. Various environmental factors, however, do play an important role since even in identical twins the correlations are only about one-half of what would be expected with 100 % genetic determination of the trait. All these data are consistent with a hypothesis of multifactorial inheritance that postulates the operation of multiple undefined genes as well as the action of various environmental mechanisms.

Ecogenetics of Hypertension – Ethnic Differences

Blood pressure rises with age. An environmental factor that often has been implicated is high salt intake. The large Intersalt study (Intersalt Cooperative Research Group 1988; Stamler et al. 1989) that correlated blood pressure and salt intake in many different populations showed significant correlation when underdeveloped populations with extremely low salt intake were included. Omitting these populations, the correlations within the higher ranges of salt intake and blood pressure were small to nonexistent. The explanation for these findings may come from the following considerations. A significant proportion of the population is "salt sensitive" and responds with increased blood pressure to salt intake while others are "salt resistant" (Luft et al. 1987). When blood pressure responses to salt are plotted, a graded distribution is obtained and no clear distinction between salt sensitivity and salt resistance is apparent. Salt sensitivity therefore has to be defined somewhat arbitrarily and 25%–50% of the population may be salt sensitive. There are strong suggestions that this response may be genetically controlled (Luft et al. 1987) but the mechanisms are unknown. The unimpressive results of the Intersalt study regarding salt intake and blood pressure in developed countries, where salt intake is high, may relate to the fact that a large proportion of the population does not raise its blood pressure with dietary salt ingestion.

Hypertension does not exist or is extremely rare in several different underdeveloped populations in various parts of the world (Table 1). When populations without hypertension migrate to environments that are more similar to those of developed countries, a certain proportion of these populations will develop hypertension (Table 1). It is likely that the genes predisposing to hypertension exist in these populations but that the appropriate environmental determinants that allow expression of these genes are absent in their original habitat. Besides increased salt intake, such environmental factors include lower potassium and calcium ingestion, higher alcohol consumption, more obesity, increased stress, and less physical exertion (see Burke and Motulsky 1991 for discussion). Genes predisposing to hypertension exist but are not expressed until

Table 1. Hypertension in underdeveloped populations

Populations with minimal hypertension[a]
Africa
Pygmies (Central Africa)
Nomads (Kenya and Uganda)
Bushman (Botswana)
South Pacific
Malaysian islanders (Western Malaysia)
Micronesians (Ontung, Java)
Melanesians (New Guinea, Solomon Islands)
Easter Islanders (Easter Island)
Polynesians (Caroline Islands, Fiji, Cook Islands)
Australia
Aborigines
South America
Aymara (Chile)
Amerindians (Guatemala, Brazil, Surinam, Mexico)
Greenland
Eskimos
China
Chinese aborigines (Szechwan)
Populations with increased blood pressure after migration or acculturation to western society
Melanesians (Solomon Islands)
Sumburu nomads (Kenya)
Tokelanau islanders (New Zealand)
Polynesians (Caroline Islands)
Zulus (Africa)

[a] From Page 1978

exposed to the appropriate environmental factors. Therefore, hypertension can be considered as an ecogenetic trait in that genes alone do not produce the untoward effect unless interaction with certain environmental factors occurs.

The mean blood pressure level of the US Black population is higher than that of US Caucasoids (Kleinbaum et al. 1971). Consequently, the number of individuals defined as hypertensive is about twice as high among Blacks as among Caucasoids. The higher frequency of medical complications among Black is consistent with the higher frequency of hypertension in this group. While the explanation of this finding could be either environmental or genetic in origin, some gene-environmental interaction is most likely. Inferential findings that suggest a genetic effect include a correlation of darkened skin color and elevated blood pressure (Darlu et al. 1990) which, however, has not been found in all studies (Hutchinson 1986). The high frequency of the genetically determined increased Na-Li countertransport abnormalities (see below) in whites and not in blacks with hypertension also is in favor of genetic effects (Motulsky et al. 1987). More critical studies are required to elucidate the relative roles of genes and environment in explaining the racial differences in blood pressure.

Genetic Models and Genetic Approaches

With multifactorial inheritance an undefined but large number of genes, each contributing a small quantitative effect, is postulated to interact with various environmental factors such as diet, stress, obesity, and others. The conventional analysis of multifactorial-polygenic mechanisms is statistical and cannot provide information regarding the number and mechanisms of the involved genes. Gene action remains a "black box." Genetic heterogeneity, i.e., different sets of genes operative in different families, makes analysis difficult and interpretation hazardous.

Some genes may have a large quantitative effect on blood pressure and a fairly large proportion of the variability of blood pressure may be caused by only a few genes. A specific search for the biologic action of such postulated major genes, therefore, has a high priority. It can be shown by simple modelling for a quantitative graded trait that only a few genes can account for a bell-shaped distribution curve. A model that posits a few major genes acting together with polygenes of small effect is therefore realistic. In the search for major genes, attention needs to be given to pathophysiologic phenomena that are associated with raised blood pressure rather than to blood pressure itself.

The search for major genes requires family studies of the trait selected for study. Computerized methods for complex segregation analysis exist to test how the obtained segregation data fit a variety of models of genetic and nongenetic familial transmission (Lalouel et al. 1983). When such methods yield data favoring models that include a major single gene, further biologic studies are highly desirable. In statistical analysis, nonbiologic artefacts such as skewing of a quantitative distribution to higher levels may mimic major gene action statistically. The study of genetic linkage is a methodology that may help to solidify the evidence for major single gene action suggested by segregation analysis.

Genetic linkage studies have already been successfully applied to a variety of clear-cut monogenic conditions. Attention is now being given to studies of genetic linkage for complex traits and diseases (Risch et al. 1990). In these approaches a chromosomal site that is tagged by a DNA variant (the marker gene) is traced in pedigrees with the genetic disease. Cosegregation or "hitchhiking" of the marker gene with the disease gene under study indicates genetic linkage. The statistical methodology to prove linkage has grown complex. Various computer programs are available but need to be applied with great caution by genetically and statistically unsophisticated users. Ideally, linkage studies should be initiated with marker genes that have a high biologic plausibility of being causally involved in the disease or trait, i.e., candidate genes. If no prior pathophysiologic knowledge exists, utilization of many anonymous markers spread evenly over all the chromosomes may lead to gene localization, particularly as the genomic map is becoming saturated with DNA markers. The results of chromosomal localization followed by gene isolation in some instances have been spectacular for clear-cut single gene disease. Results with complex disease have been less clear. Claims for certain linkages in schizophrenia and manic depressive psychosis could not be replicated (see Risch et al. 1990). Whether the problems relate to genetic heterogeneity, fuzzy phenotype assignments, absence of intermediate phenotypes, or problems with the analysis is not yet clear. All these factors may have contributed to the current impasse. No such studies have yet been done with hypertension (but see below). Attempts to study hypertension by linkage will be difficult since clear-cut monogenic transmission of hypertension rarely, if ever, occurs. Attention to intermediate phenotypes is likely to be more successful than using the trait (i.e., hypertension) itself.

Model Studies with an Intermediate Phenotype: Na-Li Countertransport

Various studies have shown that red cell Na-Li countertransport is elevated in many but not all hypertensives of Caucasoid origin (see Turner et al. 1985 for references). The exact physiologic significance of this finding is not known but presumably relates to yet undefined ion exchange abnormalities in smooth muslce cells and/or renal cells that predispose to hypertension. Various family studies have shown that countertransport values demonstrate familial aggregation (see Motulsky et al. 1987). MZ twins are most similar while spouses show little resemblance. Complex segregation analysis of countertransport has been done by our group and by other suggesting the operation of a major single gene (Motulsky et al. 1987). In our work, this gene contributed 40 % of countertransport variability while 40 % was contributed by polygenes and 20 % by environmental factors (unpublished).

The Na-H antiporter system was thought to be a likely candidate to explain variation of Na-Li countertransport. A linkage study of Na-Li countertransport with a DNA variant of the Na-H antiporter gene, however, failed to show cosegregation of these traits and appears to rule out variation of the Na-H antiporter system to explain the Na-Li countertransport results (Lifton et al. 1991). Further work with other candidate genes or linkage studies with anonymous DNA markers to search for chromosomal location of the putative major gene are required.

Our data suggest that the postulated major gene affecting countertransport contributes between 14% and 19% of the attributable risk of systolic hypertension (Motulsky et al. 1987). If this work can be replicated, a significant proportion of the variance for blood pressure has been accounted for – a promising result. This approach illustrates how a physiologically abnormal finding, i.e., elevated countertransport in hypertension, was studied as an intermediate phenotype in family investigations. Complex segregation analysis suggested the operation of a single major gene that contributes significantly to hypertension. Further genetic studies need to be done.

Animal Models

Animal models for genetic hypertension have been extensively used. While useful and interesting in demonstrating mechanisms of high blood pressure, caution must be used in transferring results to human hypertension. Genes operative in rodent hypertension may not be relevant for human hypertension. Certain genes selected for by many generations of selective breeding may only play a small or no role in human hypertension. Furthermore, the considerable species differences in the genetic background between humans and rodents may lead to quite different gene expression. However, once a specific gene has been localized in a rodent model, the DNA probe for this gene can be utilized in families with human hypertension to elucidate whether variation at the genetic locus may also be operative in human hypertension. Animals who are "outliers" for hypertension can be bred and used for linkage studies once a DNA marker map for the entire set of chromosomes is available for the species under study. Methods of linkage analysis have been developed to assign chromosomal locations to several loci (rather than to a single locus) that are implicated in a given trait (Lander and Botstein 1989). This approach has been successful for a variety of commercially useful traits in tomatoes (Paterson et al. 1988) and is being applied to the study of rat hypertension. Unfortunately, this type of linkage work is not feasible in humans since the required matings between extreme outliers do not exist.

References

Burke W, Motulsky AG (1991) Hypertension. In: King RA, Rotter J, Motulsky AG (eds) The genetic basis of common disease. Oxford, New York, in press

Darlu P, Sagnier PP, Bois E (1990) Genealogical and genetical African admixture estimations, blood pressure and hypertension in a Caribbean community. Ann Hum Biol 17: 387–397

Hutchinson J (1986) Relationship between African admixture and blood pressure variation in the Caribbean. Human Hered 36: 12–18

Intersalt Cooperative Research Group (1988) Intersalt: an international study of electrolyte excretion and blood pressure. Results for 24-hour urinary sodium and potassium excretion. Br Med J 297: 319–328

Kleinbaum DG, Kuepper LL, Cassel JC, Tyroler HA (1971) Multivariate analysis of risk of coronary heart disease in Evans County, Georgia. Arch Intern Med 128: 943–948

Lalouel JM, Rao DC, Morton NE, Elston RC (1983) A unified model for complex segregation analysis. Am J Human Genet 35: 816–826

Lander ES, Botstein D (1989) Mapping Mendelian factors underlying quantitative traits using RFLP linkage maps. Genetics 121: 185–199

Lifton RP, Hunt SC, Williams RR, Pouyssegur J, Lalouel J-M (1991) Exclusion of the Na^+-H^+ antiporter as a candidate gene in human essential hypertension. Hypertension 17: 8–14

Luft FC, Miller JZ, Weinberger MH, Grim CE, Daugherty SA, Christian JC (1987) Influence of genetic variance on sodium sensitivity of blood pressure. Klin Wochenschr 65: 101–109

Motulsky AG, Burke W, Billings PR, Ward RH (1987) Hypertension and the genetics of red cell membrane abnormalities. In: Boch G, Collins GM (eds) Molecular approaches to human polygenic disease (Ciba Foundation Symposium 130). John Wiley & Sons, Chichester, UK, pp 150–166

Page LB (1978) Hypertension and atherosclerosis in primitive and acculturated societies. In: Hypertension update. Health Learning Systms, Blumfield, pp 1–11

Paterson AH, Lander ES, Hewitt JD, Peterson S, Lincoln SE, Tanksley SD (1988) Resolution of quantitative traits into Mendelian factors by using a complete RFLP linkage map. Nature 335: 721–726

Risch N, Elston RC, Wilson AF, Gershon ES, Green P, Matthysse S, Morton NE, Ott J, Suarez BK, Reich T, Rice JP, Cloninger CR (1990) Genetic linkage and complex diseases, with special reference to psychiatric disorders. Genet Epidemiol 7: 3–46

Stamler J, Rose G, Stamle R, Elliott P, Dyer A, Marmot M (1989) Intersalt study findings. Public health and medical care implications. Hypertension 14: 570–577

Turner ST, Johnson M, Boerwinkle E, Richelson E, Taswell HF, Sing CR (1985) Sodium-lithium countertransport and blood pressure in healthy blood donors. Hypertension 7: 955–962

Ward R (1990) Familial aggregation and genetic epidemiology of blood pressure. In: Laragh JH, Brenner BM (eds) Hypertension: pathophysiology, diagnosis, and management. Raven, New York, pp 81–100

A Population Perspective for Genetics Research and Applications to Control Cardiovascular Disease in Utah*

R. R. Williams, S. C. Hunt, S. J. Hasstedt, P. N. Hopkins, L. L. Wu, M. C. Schumacher, T. D. Berry, B. M. Stults, G. K. Barlow, R. P. Lifton, and *J. M. Lalouel*

Summary

A sample of high risk families with early coronary heart disease (CHD) and hypertension (HBP) has been objectively ascertained from the general Utah population for both research efforts and public health applications regarding genetic predisposition to cardiovascular disease. Death certificates linked to population genealogical data were used to ascertain large multigenerational pedigrees. Detailed "health family tree" questionnaires obtained from the parents of 40 000 high school students have been used more recently to ascertain sibships with two or more affected siblings with either early hypertension or early coronary heart disease. Detailed biochemical and genetic studies have been performed for 4000 participants from 2500 families evaluated over the past 14 years at the Cardiovascular Genetics Research Clinic.

For both coronary heart disease and hypertension, familial aggregation is strong and due mostly to genetic factors rather than shared family environment. As expected, lipid abnormalities were up to five times more common among coronary siblings than in the general population, but hypertensive siblings also showed up to four times more frequent lipid abnormalities than the general population. Familial dyslipidemic hypertension (FDH) was defined as two or more siblings having both hypertension and lipid abnormalities before age 60. About 12 % of persons with hypertension have this disorder and also have increased risk of coronary heart disease. Persons with FDH often have elevated fasting insulin levels and either familial combined hyperlipidemia or central obesity. Persons with a point mutation for lipoprotein lipase seem to have more hypertension and lipid abnormalities and may help explain a subset of FDH. Understanding and diagnosing FDH has practical application since it leads to better diagnosis and treatment of lipid abnormalities both in index cases with hypertension and in their close relatives who share their risk.

* Supported by two grants (HL24855-11 and HL21088-14) from the National Heart, Lung and Blood Institute, Bethesda, MD, USA and a cooperative agreement with the U.S. Centers for Disease Control, Atlanta, GA, USA. Public helath applications to control cardiovascular disease in high risk families is a joint project with the Bureau of Chronic Disease Control of the Utah Department of Health

K. Berg, V. Bulyzhenkov, Y. Christen, P. Corvol (Eds.)
Genetic Approaches to Coronary Heart Disease
and Hypertension

Segregation analysis in hypertensive families has detected seven recessive "intermediate phenotypes" with high heritability and some association with hypertension. These include urinary kallikrein excretion, sodium-lithium countertransport, intraerythrocytic sodium concentration, number of erythrocytic sites of the sodium potassium ATPase pump, a relative fat pattern index, knee width, and possibly fasting plasma insulin levels. A seven year prospective study found several variables were independent predictors of the future occurrence of hypertension, including the high sodium-lithium countertransport genotype, a positive family history, uric acid, phosphate, systolic blood pressure, change in diastolic blood pressure during bicycle exercise, change in diastolic blood pressure during handgrip exercise, and scapular skinfold thickness.

Three approaches to genetic linkage studies have been illustrated by a test for genetic linkage of recessive high sodium-lithium countertransport with the antiporter gene of the sodium hydrogen exchange system. No cosegregation was found in extended pedigrees, no increased sharing of alleles was found between hypertensive siblings, and there was no correlation between similar SLC levels in siblings and increased sharing of alleles. These three approaches will be used in further tests for genetic linkage of traits related to hypertension.

Characterization of families with two or more living siblings with early CHD revealed several common syndromes: familial combined hyperlipidemia (FCHL) in 36%, familial dyslipidemic hypertension (FDH) in 21%, low high density lipoprotein (HDL)-cholesterol in 15% (with considerable overlap of FDH with FCHL and low HDL); high Lp(a) in 16%, high homocyst(e)ine in 6%, familial hypercholesterolemia (FH) in 3% and type III hyperlipidemia in 3%.

A prototype for helping high risk families is sponsored by the U.S. Centers for Disease Control and has been given the title MED PED (Tracing *Med*ical *Ped*igrees to Foster *M*ore *E*arly *D*iagnoses and *P*revent *E*arly *D*eaths). The goal is to find and help all persons in Utah with FH. Index cases with confirmed FH are collected from physicians and laboratories, high risk relatives are identified and screened for FH, and a computer registry tracks the results of screening and treatment. Coordinated efforts in population genetics can lead to effective research and help initiate public health measures to help families with high risk of cardiovascular disease.

Introduction

For the past 14 years, the University of Utah Cardiovascular Genetics Research Clinic has pursued the goal of gaining a population perspective for both research and application regarding genetic factors leading to cardiovascular disease. We are finding that a large proportion of early cardiovascular disease involves strong genetic predispositions. We are identifying specific genetic and pathophysiological mechanisms and trying to estimate the proportion of each mechanism contributing to the disease in the general population. We are also interested in understanding the interactions between different genetic factors and environmental factors. We are working with public health professionals to pilot projects that illustrate how this new knowledge can be applied to help control genetically predisposed cardiovascular disease.

Methods

From 1976 to 1980, a computerized genealogical database for the Utah population was developed and linked to death certificates to estimate the proportion of cardiovascular disease occurring in family aggregations and to help ascertain population-based high risk families (Williams et al. 1979; Williams 1980). From 1980 to 1985, 2,500 individuals in 98 families were ascertained, either

1. two or more coronary siblings dying before age 55 of CHD;
2. two or more siblings dying of stroke before age 70; or
3. index cases from the Hypertension Detection and Follow-up Program of the National Heart, Lung and Blood Institute (Williams and Hunt 1987). From 1985 to 1990, an additional 150 families with two or more hypertensive siblings and 22 families with two or more coronary siblings at an early age were ascertained from "Health Family Tree" questionnaires collected from the parents of 40,000 high school students (Williams et al. 1988a).
4. The occurrence of hypertension and coronary disease was validated by mail and phone contact before approximately 1,500 relatives from these families were invited for a detailed research clinic evaluation.

Variables collected during this study protocol included: lipids, lipoproteins, and apolipoproteins (Wu et al. 1989), plasma and urine electrolytes and blood chemistries; intracellular cations and membrane cation transport tests (Smith et al. 1984); a battery of 28 blood pressures and multiple anthropometric measurements; detailed physician's history and physical examination and personal health questionnaires; and white blood cells stored for future DNA extraction and molecular biological studies.

Statistical analyses included standard epidemiological methods (Williams et al. 1983), complex maximum likelihood segregation analysis testing for major gene effects and estimating both major gene and polygenic heritability (Hasstedt et al. 1988a), and genetic linkage analysis following three approaches: cosegregation in multigenerational pedigrees, testing for excess sharing of alleles among affected sibling pairs in sibships with hypertension, and testing for correlation between the similarity of quantitative values of a physiologic trait in siblings versus the number of alleles shared at a marker locus (Lifton et al. 1990a).

Whenever the genotype status of individuals can be inferred from segregation analysis, linkage analysis, or evaluation of point mutations, persons with contrasting genotypes were characterized with respect to clinical variables such as hypertension and lipid abnormalities to detect the effects of specific genetic loci on observed cardiovascular disease and its risk factors (Hasstedt et al. 1988a; Wilson et al. 1990).

In prototypes of disease control, high risk families are indexed from family history questionnaires (Health Family Trees; Hunt et al. 1986), computerized records of patients with early coronary disease in hospitals, striking abnormalities in lipid levels from laboratories, and from mail questionnaires sent to all practicing primary care physicians in Utah. Beginning with a computerized register of persons with confirmed familial hypercholesterolemia (FH), mail and phone contacts with these index cases and their close relatives allow a sequential expansion of pedigrees to find other relatives who also have FH. A growing computerized register of families with FH is being used to try to find all persons with FH in Utah and to track their progress regarding treatment and control of cardiovascular disease in the future.

Results

Familial Aggregation of Hypertension and Early Coronary Disease

As shown in Table 1, moderate increased risk for future hypertension and coronary disease was observed in persons with a mild positive family history (defined as having one or more first-degree relatives affected before age 55). A more selective definition (two or more affected first-degree relatives under age

Table 1. Cardiovascular disease family history: frequency and relative risk[a] (Adapted from Hunt et al. 1986)

Definition of a positive[b] family history	Hypertension		Coronary disease	
	Frequency	Relative risk	Frequency	Relative risk
1 + affected < age 55	32%	2.8	13%	3.9
2 + affected < age 55	11%	4.1	2%	12.7

[a] Data from males aged 20–39 years in 7625 Utah families
[b] Number of first degree relatives affected before age 55

Table 2. Dyslipidemia in siblings with early CHD and hypertension (HBP)[a] (Adapted from Williams et al. 1988b, 1990a)

	Ratio of observed to expected frequencies of extreme values					
Siblings studied	*LDL-cholesterol* percentiles		*Triglycerides* percentiles		*HDL-cholesterol* percentiles	
	90th	95th	90th	95th	10th	5th
HBP[b]	1.9	2.6	3.0	3.4	3.9	4.2
CHD[c]	2.0	2.7	3.9	5.5	3.6	5.0

All observed expected ratios are significantly greater than 1.0 at $P <$
[a] Expected values from age- and sex-specific LRC percentile data
[b] N = 133 siblings with HBP < age 60
[c] N = 44 siblings with CHD < age 55

55) is much less common but it identifies a subset of individuals with very high risk of future hypertension and coronary disease (Hunt et al. 1986).

As shown in Table 2, siblings with hypertension before age 60 and siblings with coronary disease before age 55 have lipid abnormalities up to four or five times more often than expected in the general population. It is interesting to note that siblings identified only for multiple occurrence of hypertension in their sibship have lipid abnormalities almost as striking as those found in two or more siblings with coronary disease before age 55 (Williams et al. 1989b, 1990a).

Familial Dyslipidemic Hypertension – A Subset of Essential Hypertension

A new syndrome, familial dyslipidemic hypertension (FDH), was defined as an observation of two or more siblings who both have hypertension and lipid abnormalities diagnosed before age 60. These lipid abnormalities include HDL-cholesterol below the 10th percentile, triglycerides above the 90th percentile, or LDL-cholesterol above the 90th percentile (Williams et al. 1988b). Many persons with FDH have increased fasting plasma insulin levels, increased apolipoprotein B levels, and either familial combined hyperlipidemia or central obesity (Hunt et al. 1989). Persons with FDH often have a strong positive family history for coronary disease, and among families with early coronary disease, FDH is one of the most common syndromes found (Williams et al. 1990). FDH seems to occur in about 1 % of the general population, 12 % of persons with hypertension, and 21 % of persons with early familial coronary heart disease (Williams et al. 1988b, 1990a). Similar frequencies for occurrence of FDH were observed in a study in the NHLBI twin study (Selby et al. 1989). This study helps to verify that FDH is not a result of medications, since 80 % of participants in that study were evaluated before any medictions had been

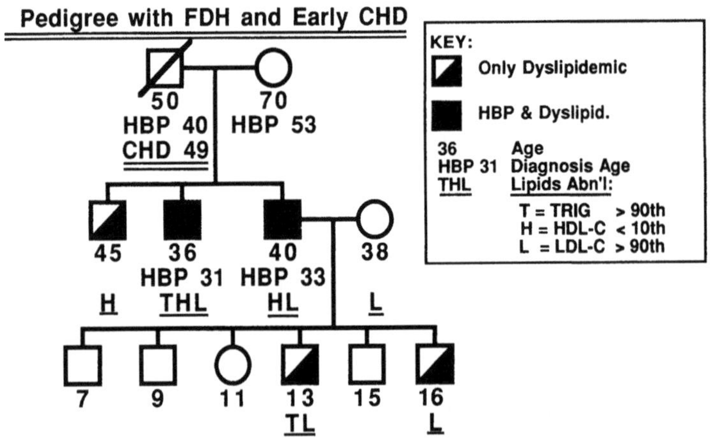

Fig. 1. Pedigree showing early hypertension (HBP) and early coronary disease (CHD). Criteria are met for diagnosing both familial combined hyperlipidemia (FCHL) and familial dyslipidemic hypertension (FDH). (Part of a previously published large pedigree, Williams et al. 1990a)

initiated. Potential genetic contribution to FDH was also suggested by monozygous twins having FDH concordance three times more often than dizygous twins (Selby et al. 1989). A family with FDH is illustrated in Fig. 1. Among several large pedigrees with FDH, it would appear that dyslipidemia occurs 10–20 years before hypertension. Dyslipidemia also occurs in some relatives without hypertension. Coronary events occur about 10 years after hypertension. Recent data suggests that a point mutation causing lipoprotein lipase deficiency may be related to FDH (Wilson et al. 1990). While no effect was observed in those who carried the mutation under age 40, those who are 40 years or older showed several significant observations: 58 % had triglyceride levels above the 95th percentile and HDL-cholesterol levels below the 5th percentile and 68 % met standard criteria for the diagnosis of hypertension (Wilson et al. 1990).

While the genetic and pathophysiological mechanisms of FDH remained to be completely defined, current understanding of this syndrome already suggests practical approaches to controlling familial cardiovascular disease. Younger hypertensives with positive family history should be evaluated for cholesterol, triglyceride and HDL levels. Persons found to have FDH should be carefully followed for both hypertension and lipid abnormalities, and their close relatives should be checked for similar manifestations.

Studies in many countries have suggested an important role for insulin resistance and hyperinsulinemia in the pathophysiology of hypertension. Figure 2 illustrates a pathophysiological model that could relate genetic and

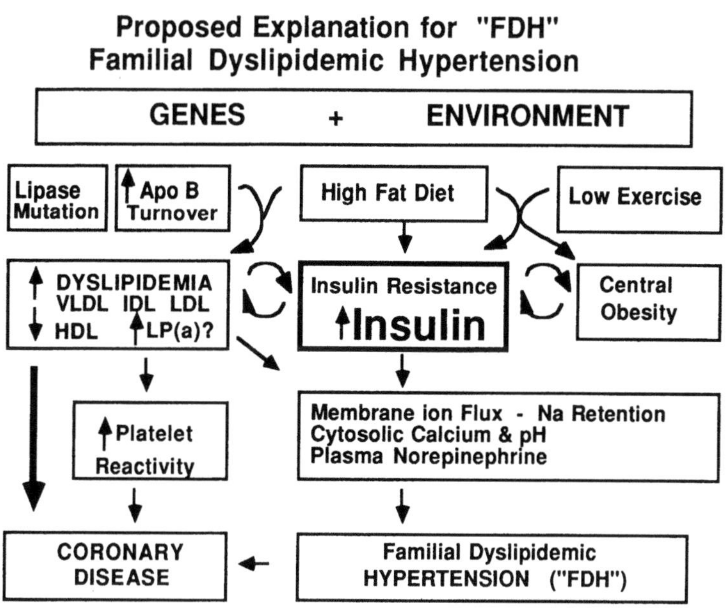

Fig. 2. Theoretical model of pathophysiology. The model suggests that genetic traits (such as lipoprotein lipase deficiency or high apolipoprotein(b) (Apo (b)) turnover rate) combine with environmental factors (such as high fat diet or low exercise level) to foster dyslipidemia, obesity, and insulin resistance. The final effect is high risk for FDH and early CHD

environmental factors that could contribute to the development of both hypertension and early coronary disease involving mechanisms that relate insulin resistance to lipid abnormalities and obesity (Kaplan et al. 1989).

Complex segregation analysis using maximum likelihood methods has found evidence for recessive segregating major gene effects and very high estimates of heritability for several factors potentially related to FDH. These include a relative fat pattern index (the ratio of scapular skinfold thickness to the sum of scapular and suprailiac skinfolds; Hasstedt et al. 1988a); knee width, and fasting insulin levels. (The latter two observations are recent unpublished results from the Utah Cardiovascular Genetics Research Clinic).

Highly Heritable Cation Traits and Hypertension

Most popular theories regarding the pathophysiology of hypertension propose a subset of the population have a genetic susceptibility to hypertension that is promoted by adverse dietary electrolyte intake (high sodium, low potassium, etc.). Some proposed "intermediate phenotypes" in pathophysiological models of hypertension involving cations have been found to be highly heritable genetic traits with significant recessive segregation and include urinary kallikrein excretion (Berry et al. 1989), sodium-lithium countertransport (SLC; Hasstedt et al. 1988a), intraerythrocytic sodium concentration (Hasstedt et al. 1988b), and the number of erythrocytic pumping sites of the sodium potassium ATPase pump (Hasstedt et al. 1989b). Total heritability estimates for these traits range from 77% to 84% and generally include both major gene and polygenic contributions. Probably the best studied of these traits is sodium-lithium countertransport, which has been shown to have strong genetic and potential major gene effects in three different populations (Hasstedt et al. 1988a; Boerwinkle et al. 1984; Motulsky et al. 1987) and was a prospective predictor of future hypertension in a seven-year follow-up study (Hunt et al. 1990). It is thought that this non-physiologic laboratory test reflects the sodium hydrogen exchange system which does operate physiologically (Semplicini et al. 1989). The gene for the structural antiporter protein involved in this transport system was cloned in France (Sardet et al. 1989). To test the hypothesis that the antiporter locus is linked with sodium-lithium countertransport, we developed polymorphic markers at this locus (Lifton et al. 1990), and tested for genetic linkage using three approaches. We failed to find cosegregation between the recessive SLC genotype from segregation analysis and the DNA polymorphisms for the antiporter locus; we failed to find increased sharing of the alleles at the antiporter locus between multiple siblings with hypertension; we also failed to find any correlation between the similarity of SLC levels in siblings and the number of alleles shared at the antiporter locus (Lifton et al. 1991). While this study failed to find genetic linkage, it provides a good example of the three major approaches for testing genetic linkage in future studies of the genetics of hypertension.

Genetic Predisposition to Early Coronary Heart Disease

Table 3 lists the frequency of seven metabolic syndromes found in siblings with two or more living CHD cases ascertained from Health Family Trees in Utah. Familial combined hyperlipidemia (FCHL) was the most common syndrome followed by FDH and low HDL-cholesterol levels. Low HDL-cholesterol and high triglycerides are the most common manifestations of all three of these syndromes. This helps to explain why siblings with early CHD show triglyceride and HDL-cholesterol abnormalities more often than LDL-cholesterol elevations, as shown in Table 2. As noted at the bottom of Table 3, there is some overlap, since most persons with FDH have either FCHL or low HDL-cholesterol syndromes. It is worth noting that the category of "low HDL-cholesterol" includes only coronary sibships in which both siblings have normal cholesterol but low HDL-cholesterol as well as coronary disease before age 55. This emphazises the importance of screening triglyceride and HDL levels as well as total cholesterol in evaluating persons with a positive family history of early CHD.

High levels of Lp(a) are highly heritable (Hasstedt et al. 1983), genetically linked to the plasminogen gene (Drayna et al. 1988), and commonly observed in multiplex sibships with early CHD. This common risk factor deserves more attention if it is to become a clinically useful marker of genetic predispositions to early CHD.

Homocyst(e)ine levels in the bloodstream have recently been identified to occur significantly more often than expected in siblings with early CHD (Williams et al. 1990b). It also correlates highly in siblings but not spouses suggesting genetic determination (Williams et al. 1990b), and is thought to be a trait that would respond quite easily to simple dietary supplementation with folic acid (Malinow 1990). Other studies have confirmed the association of hyperhomocyst(e)inemia with early coronary disease in Boston (Genest et al. 1990) and South Africa (unpublished results communicated by Professor Devilliers, Department of Chemical Pathology, University of Pretoria).

Table 3. Utah families with early CHD (Adapted from Williams et al. 1990a, b)

Syndrome	*Early CHD* (%)	*Prevalence* (%)
1. FCHL	36	2
2. FDH[a]	21	1
3. Low HDL-cholesterol	15	1
4. High Lp(a)	16	1
5. Homocyst(e)ine	6	1
6. FH	3	0.2
7. Type III hyperlipidemia	3	0.2

[a] Most persons with FDH also have either FCHL or low HDL

MED PED – A Prototype for Controlling Genetically Determined Cardiovascular Disease

While FH is one of the least common syndromes in Table 3, it is probably the best understood. Goldstein and Brown received the Nobel Prize in 1985 for their lucid observations regarding defects in LDL receptors as a cause of this disorder. The gene for this condition has now been mapped to chromosome 19 (Leppert et al. 1988), and studies now suggest that a combination of dietary and drug therapy should dramatically delay the onset of early atherosclerosis in heterozygotes with this condition (Kane et al. 1990).

FH is currently about as frequent as AIDS, about as lethal as AIDS and, at the present time, much more treatable than AIDS. While we applaud the success of our colleagues in research and public health regarding AIDS, we should follow their example in trying to raise public awareness and support for programs to find and help persons with FH to prevent early coronary deaths.

As a pilot effort in this regard, the U.S. Centers for Disease Control has funded project MED PED (Tracing Medical Pedigrees to Foster More Early Diagnoses and Prevent Early Deaths). The first step is to collect a register of index cases with confirmed FH ascertained from lipid specialists and other physicians canvassed using a letter sent to all primary care physicians in the Utah State Medical Association. Other index cases with FH are being ascertained from computer monitoring of extremely high LDL-cholesterol levels in computer files of hospitals and major laboratories. MED PED staff contact these index cases to identify first- and second degree relatives who have a 25 % to 50 % chance of also having familial hypercholesterolemia. Further sequential testing of relatives follows any individuals found to have this diagnosis. In the third step, a computer registry of FH pedigrees is built to track results of screening and treatment and to coordinate further efforts to find and help more individuals with FH.

As a fourth step in Utah, a computerized genealogical database containing all families has been searched to identify ancestors for several independently ascertained pedigree showing that, several generations farther up their family tree, they converge into a single large pedigree. In one case, 11 four generation pedigrees with FH have converged into a single, large seven-generation pedigree. We are now following genealogical records to find many descending branches of the large seven generation pedigree. We estimate that this single pedigree will contain 2000 persons with familial hypercholesterolemia. Persons from this pedigree are already spread across ten different states in the United States showing, the need for broad population coordination for such public health efforts.

It is hoped that project MED PED will show the cost efficiency of this pedigree tracing approach for controlling a well-defined genetic predisposition to early coronary heart disease. One of the objectives of this project is to develop methods (e.g., personal computer programs) that would be easily adapted to other locations. We hope that a genetically focused public health effort such as MED PED could expand under the auspices of the World Health Organization.

While such genealogical efforts may seem to be uniquely feasible only in Utah, a rapidly expanding set of tools for tracing, storing and sharing ancestries is expanding throughout the world. A joint effort of the Church of Jesus Christ of Latter Day Saints (Mormons), working together with genealogical associations in several countries, is producing a large international collection of ancestral data. A network of 1500 family history centers in major cities throughout the world have access to microfilms containing vital information on two billion deceased persons living during the years 1500–1900. For some locations in the British Isles and Scandinavia, almost all records of births, marriages and deaths during that time period have been microfilmed and catalogued in this collection. A collection of 10 000 microfiche available in each of these 1500 family history centers contains a complete catalog of all microfilms and books in the central collection as well as a growing international genealogical index of names of deceased persons with their spouses and parents. Within the past year many of these family history centers have acquired the computer hardware and software necessary to access a growing collection of compact discs containing vital records such as the following:

- 147 million deceased persons with their spouses or parents.
- 39 million name U.S. social security death register.
- 15 million name 1880 U.S. census.
- 27 million name English 1881 census.
- 5 million name file of Australian births, marriages and deaths.
- "Ancestral file" a computerized lineage-linked database with 7 million names currently and projected to grow to include the shared ancestries available for all mankind.

One geneticist suggested this genealogical effort seems to be an ancestral parallel to the human genome project. Neither of these two massive worldwide efforts is trivial but both seem to have potential benefits for both research and public health efforts relating to genetic factors and cardiovascular disease.

Summary Recommendations

First, is recommended that standard methods and syndrome definitions be adopted to facilitate comparable ascertainment, screening evaluations, definitions of positive family history and assessment of syndromes such as FCHL and high levels of Lp(a).

Second, is recommended that international comparison of comparable data be undertaken to contrast the frequencies of syndromes and the progress in different countries regarding the diagnosis and treatment of genetic syndromes for cardiovascular disease (parallel to efforts already undertaken by epidemiologists regarding the incidence and mortality of coronary disease).

Third, is suggested that the World Health Organization foster an effort to coordinate and prioritize actions for controlling genetically determined cardiovascular diseases. Parallel efforts for finding and helping high risk families could be initiated in several different countries. A World Health Organization registry could be initiated for high risk pedigrees.

Fourth, is suggested that the practical knowledge regarding some syndromes such as FH and FDH is sufficient to begin prototype projects as true genetic approaches for controlling hypertension and early coronary heart disease.

References

Berry TD, Hasstedt SJ, Hunt SC, Wu LL, Smith JB, Ash KO, Kuida H, Williams RR (1989) A gene for high urinary kallikrein may protect against hypertension in Utah kindreds. Hypertension 13: 3–8

Boerwinkle E, Turner ST, Sing CF (1984) The role of the genetics of sodium lithium countertransport in the determination of blood pressure variability in the population at large. In: Brewer GS (ed) The red cell: sixth Ann Arbor conference. Liss, New York, pp 479–507

Drayna DT, Hegele RA, Haas P, Emi M, Wu LL, Eaton DL, Lawn RM, Williams RR, White RL, Lalouel JM (1988) Genetic linkage between lipoprotein (a) phenotype and a DNA polymorphism in the plasminogen gene. Genomics 3: 230–236

Genest J, McNamera JR, Salem DN, Wilson PWF, Schaefer EJ, Malinow MR (1990) Plasma homocyst(e)ine levels in men with premature coronary artery disease. J Am Coll Cardiol 16: 1114–1119

Hasstedt SJ, Wilson DE, Edwards CQ, Cannon WN, Carmelli D, Williams RR (1983) The genetics of quantitative plasma Lp(a): analysis of a large pedigree. Am J Med Genet 16: 179–188

Hasstedt SJ, Wu LL, Ash KO, Kuida H, Williams RR (1988a) Hypertension and sodium-lithium countertransport in Utah pedigrees: Evidence for major locus inheritance. Am J Hum Genet 43: 14–22

Hasstedt SJ, Hunt SC, Wu LL, Williams RR (1988b) The inheritance of intraerythrocytic sodium level. Am J Med Genet 29: 193–203

Hasstedt SJ, Ramirez ME, Kuida H, Williams RR (1988a) Recessive inheritance of a relative fat pattern. Am J Hum Genet 45: 917–925

Hasstedt SJ, Wu LL, Kuida H, Williams RR (1989b) Recessive inheritance of a high number of sodium pump sites. Am J Med Genet 34: 332–337

Hunt SC, Williams RR, Barlow GK (1986) A comparison of positive family history definitions for defining risk of future disease. J Chronic Dis 39: 809–821

Hunt SC, Wu LL, Hopkins PN, Stults BM, Kuida H, Ramirez ME, Lalouel JM, Williams RR (1989) Apolipoprotein, low density lipoprotein subfraction, and insulin associates with familial combined hyperlipidemia: study of Utah patients with familial dyslipidemic hypertension. Arteriosclerosis 9: 335–344

Hunt SC, Stephenson SH, Hopkins PN, Hasstedt SJ, Williams RR (1991) A prospective study of sodium-lithium countertransport and hypertension in Utah. Hypertension 17: 1–7

Kane JP, Malloy MJ, Ports TA, Phillips NR (1990) Regression of coronary atherosclerosis during treatment of familial hypercholesterolemia with combined drug regimens. JAMA 264: 3007–3012

Kaplan NM (1989) The deadly quartet: upper-body obesity, glucose intolerance, hypertriglyceridemia, and hypertension. Arch Intern Med 149: 1514–1520

Leppert MF, Hasstedt SJ, Holm T, O'Connell P, Wu LL, Ash KO, Williams RR, White R (1986) A DNA probe for the LDL receptor gene is tightly linked to hypercholesterolemia in a pedigree with early coronary disease. Am J Hum Genetics 39: 300–306

Lifton RP, Hunt SC, Williams RR, Pouysségur J, Lalouel JM (1991) Exclusion of the Na^+/H^+ antiporter as a candidate gene in human essential hypertension. Hypertension 17: 8–14

Lifton RP, Sardet C, Pouyssegur J, Lalouel JM (1990) Cloning of the human genomic amiloride-sensitive Na^+/H^+ antiporter gene, identification of genetic polymorphisms and localization on the genetic map of chromosome 1p. Genomics 7: 131–135

Malinow MR (1990) Hyperhomocyst(e)inemia: a common and easily reversible risk factor for occlusive atherosclerosis. Circulation 81: 2004–2006

Motulsky AG, Burke W, Billings PR, Ward RH (1987) Hypertension and the genetics of red cell membrane abnormalities. In: Bock G, Collins G (eds) Molecular approaches to human polygenic disease. Wiley, Chichester, pp 150–166 (Ciba Symposium, vol 30)

Sardet C, Franchi A, Pouyssegur J (1989) Molecular cloning, primary structure, and expression of the human growth factor-activatable Na^+/H^+ antiporter. Cell 56: 271–280

Selby JV, Newman B, Fabsitz RR, Quiroga J, Christian JC (1989) Dyslipidemic hypertension in twins. Circulation 80: II-208

Semplicini A, Canessa M, Mozzato MG, Ceolotto G, Marzola M, Buzzaccarini F, Casolino P, Pessina AC (1989) Red blood cell Na^+/H^+ and Li^+/Na^+ exchange in patients with essential hypertension. Am J Hypertens 2: 903–908

Smith JB, Ash KO, Hentschel WM, Sprowell WL, Williams RR (1984) A simplified method for simultaneously determining countertransport and cotransport in human erythrocytes. Clin Chim Acta 137: 169–177

Williams RR (1980) A population perspective for early and familial coronary heart disease. In: Proceedings of conference on human health data from defined populations. Cold Spring Harbor Laboratory, New York, pp 333–350 (Banbury report, vol 4)

Williams RR, Hunt SC (1987) Recruitment of members of high-risk Utah pedigrees. Controlled Clin Trials 8: 105S–114S

Williams RR, Skolnick M, Carmelli D, Maness AT, Hunt SC, Hasstedt SJ, Reiber GE, Jones RK (1979) Utah pedigree studies: design and preliminary data for premature male CHD deaths. In: Sing CF, Skolnick M (eds) The genetic analysis of common diseases. Liss, New York, pp 711–729

Williams RR, Hunt SC, Kuida H, Smith JB, Ash KO (1983) Sodium-lithium countertransport in erythrocytes of hypertension prone families in Utah. Am J Epidemiol 118: 338–344

Williams RR, Hunt SC, Barlow GK, Chamberlain RM, Weinberg AD, Cooper HP, Carbonari JP, Gotto AM (1988a) Health family trees: a tool for finding and helping young family members of coronary and cancer prone pedigrees in Texas and Utah. Am J Public Health 78: 1283–1286

Williams RR, Hunt SC, Hopkins PN, Stults BM, Wu LL, Hasstedt SJ, Barlow GK, Stephenson SH, Lalouel JM, Kuida H (1988b) Familial dyslipidemic hypertension.: evidence from 58 Utah families for a syndrome present in approximately 12% of patients with essential hypertension. JAMA 259: 3579–3586

Williams RR, Hopkins PN, Hunt SC, Wu LL, Hasstedt SJ, Lalouel JM, Ash KO, Stults BM, Kuida H (1990a) Population-based frequency of dyslipidemia syndromes in coronary-prone families in Utah. Arch Intern Med 150: 582–588

Williams RR, Malinow MR, Hunt SC, Upson B, Wu LL, Hopkins PN, Stults BM, Kuida H (1990b) Hyperhomocyst(e)inemia in Utah siblings with early coronary disease. Coronary Artery Dis 1: 681–685

Wilson DE, Emi M, Iverius PH, Hata A, Wu LL, Hillas E, Williams RR, Lalouel JM (1990) Phenotypic expression of heterozygous lipoprotein lipase deficiency in the extended pedigree of a proband homozygous for a missense mutation. J Clin Invest 86: 735–750

Wu LL, Warnick GR, Wu JT, Williams RR, Lalouel JM (1989) A rapid micro-scale procedure for determination of the total lipid profile. Clin Chem 35: 1486–1491

Cell Membrane Alteration in Primary Hypertension: An Evidence of Its Genomic Source

Y. V. Postnov

The study of primary hypertension[a] has now entered a new phase in which the genome is becoming the principal focus research. Movement in this promising direction requires overcoming not only technical and technological problems. It has become evident that the choice of the loci where genomic abnormalities should be searched for will largely determine to what extent our understanding of the nature of hypertension corresponds to reality. The key chain of the pathogenesis may serve as a guide for the most efficient search for gene abberations.

However, current schemas of the pathogenesis of primary hypertension are highly contradictory. The main trends in the study of hypertension originated from both clinical and experimental observations in which causal relationships between organ or system alterations and increased blood pressure (BP) were found.

However, such approaches to the evaluation of the causal factor inevitably formed a traditional view that the main cause of primary hypertension should be sought in the initial alteration of particular mechanisms maintaining systemic BP as it was demonstrated for some secondary forms. Until recently there has been a strong tendency to cast alterations in these particular systems (often described as "pressor" or "depressor" ones) in the role of the overall causal factor in primary hypertension. Guyton and his group were among the first to recognize the inadequacy of this point of view. According to Guyton's hypothesis, the kidney is the main determinant for a long-term pressure control (as a kind of "barostat" for the cardiovascular system), while particular BP regulation mechanisms – such as the renin-angiotensin system, sympathetic nervous system, etc. – play a role of feedback variables depending on renal excretory function. According to Guyton (1980), primary hypertension is caused by an initial alteration of kidney itself, resulting in a modification of renal excretory function (also called, kidney resetting) and thus shifting the set-point of the kidney barostat mentioned above. Definitely, the latter point of view is different from those which consider the kidney to be a source of humoral agents acting on the peripheral vessels, thus producing increased vascular tone and hypertension. Although we share the

[a] Here we use the term "primary hypertension" both for essential hypertension in humans and its experimental model, the spontaneous hypertension of rats of the Okamoto-Aoki strain (SHR)

K. Berg, V. Bulyzhenkov, Y. Christen, P. Corvol (Eds.)
Genetic Approaches to Coronary Heart Disease
and Hypertension

view of the central role of the kidney in chronic hypertension of any sort, there are no data linking the appearance of primary hypertension with any initial alteration of this organs excretory function. Thus, we are still left with the underlying problem: what is the initial disturbance in essential hypertension (EH)? It is not only a theoretical question but also a practical question.

Before providing arguments for our point of view concerning what we believe is the major pathogenetic basis behind the primary forms of hypertension, we should discuss the etiological concept proposed by G. F. Lang, a Russian scientist whose great contributions to the problem are completely unknown in the West due to the prolonged isolation of this country. As far back as 1922, just after the Russian Civil War, Lang suggested a concept describing EH as a consequence of chronic neuromental overload (neuromental stress). Lang developed this concept until his death in 1948, and the influence of his ideas is still very deep in this country (Lang 1950).

Considering the irregular occurrence of hypertension in individuals subjected to neuromental stress, Lang came to a conclusion about the existence of certain genetic prerequisites ("constitutional features of the CNS") affected by exposure to environmental factors. Thus, according to Lang, essential hypertension could be regarded as a kind of disease of imperfect adaptation. In fact, much can be found in common between the ideas of Lang in Russia and Pickering (1977) in England, even though they worked and thought independently.

The next step in our study of the nature of this pathology was the search for the main determinant which constitutes the basis of primary hypertension as a disease of imperfect adaptation to environmental factors. The discovery of a certain type of membranopathy in primary forms of hypertension indicated it was such a determinant. By "membranopathy," we mean the presence in hypertension of widespread (i.e., not limited to cardiovascular contractile elements) abnormalities in the ion transport systems of the cell plasma membrane and changes in its ultrastructure as well as in certain physico-chemical properties, resulting in insufficient membrane control over intracellular calcium distribution (so-called membrane defect; Postnov and Orlov 1979, 1984).

The first membrane abnormalities were discovered in the early 1970s by Jones (1973) in vascular smooth muscles of SHR (an increased membrane permeability for K^+ and Cl^-) and by our group in erythrocytes of both SHR and EH patients (Postnov et al. 1976, 1977). In these cases we were able to show a nearly two-fold increase in the ouabain-resistant component of Na^+/Na^+ exchange, which characterizes the plasma membrane "passive" permeability to Na^+. In later studies these data were supplemented by characteristics of other ion transport systems, namely $Na^+ - K^+$ cotransport (Garay and Meyer 1979) and Li^+/Na^+ countertransport (Canessa et al. 1980), which confirmed the initial conclusion about the increased permeability to univalent cations of the erythrocyte membrane in primary but not in secondary forms of hypertension. Recently, an enhanced rate of Na^+/H^+ exchange was demonstrated in platelets (Livne et al. 1987), in erythrocytes of EH patients (Orlov et al. 1989; Postnov et al. 1988) and in those of SHR (Orlov et al. 1989) as well as in culturated smooth muscle cells of SHR (Berk et al. 1989).

After these first findings, alterations of membrane handling of cell calcium were revealed in different cell subjects, such as erythrocytes, platelets, adipocytes, cardiomyocytes, vascular smooth muscle cells, synaptosomes, etc.

The major pathogenetically significant consequence of these alterations is the appearance of increased concentrations of cytoplasmic free calcium under physiological action on the mechanisms of cell calcium homeostasis (for details see Postnov and Orlov 1985).

At least three known membrane abnormalities should be considered as a source of free calcium excess in the cytoplasm of cells:

1. Increased calcium influx rate in various types of cells (which occurs in primary hypertension despite the difference in transport systems maintaining the calcium influx in different cells
2. Decrease of calcium binding by the inner surface of the plasma membrane
3. Alteration of adenosine triphosphate dependent, calmodulin-stimulated outward calcium transport

The discovery of widespread membrane alterations seems to clarify the key pathogenetic mechanisms in primary hypertension: changes in the sensitivity to various agonists and excessive contractility of vascular smooth muscles, increased neuromediator charge in sympathetic terminals, and changed sensitivity and reactivity of certain tissues. All these may be considered as the general causes of increased BP, whose stability is maintained by the subsequent shift of renal excretory function in relation to BP (i.e., by the kidney operating in modified mode and ensuring normal output under increased systemic BP and, in turn, requiring an elevated BP level).

During the last 15 years the hypothesis which considered primary hypertension to be a manifestation of a widespread membrane alteration underwent rigorous experimental testing, from which the following conclusion can be drawn:

1. The widespread alteration of plasma membrane structure and its ion transport function exists in primary hypertension and is expressed not only in cardiovascular contractile cells but also in other types of cells (This does not mean that this alteration exists *in all types* of cells in the body).
2. These membrane alterations are absent in secondary forms of hypertension.
3. The development of membrane alterations (at least in the case of the experimental model of primary hypertension, SHR) is not dependent on corticosteroid secretion (Jones 1973; Postnov et al. 1976) or on sympathetic nervous system activity (Postnov et al. 1981).
4. In the case of SHR the membrane alterations are already expressed at the prehypertensive stage and, in man, often in as yet normotensive siblings, presumably reflecting genomic sources of "membranopathy."

The characteristics of membrane alterations (their restriction to the plasma membrane per se as well as their presence in many different types of cells) began to acquire a special importance in light of data obtained in studies on erythrocytes, concerning both the involvement of the membrane skeleton and the role of protein kinase C (PKC) in the development of these membrane alterations (Postnov et al. 1987).

In erythrocytes from healthy individuals, the activation of PKC (whose activity in primary hypertension was found to be increased in different types of cells) is able to bring about changes in membrane skeleton as well as alterations of certain cation carriers similar to those observed in hypertension (Postnov et al. 1988). As PKC is an important terminal chain of the phosphoinositol lipid signaling system (PILSS) which mediates signal transduction from surface receptors to the cytoskeleton and to ion transporting systems (Berridge 1984), the above findings place this system at the center of the search for the cause of membrane alterations as well as for the etiologic basis of this pathology.

This suggestion is also supported by a study that demonstrated, more than 10 years ago, an increase of the phosphoinositide (PI) metabolism in SHR erythrocytes (Borskina et al. 1978). Alterations in PI metabolism were later also revealed in erythrocytes (Marche et al. 1985) and platelets (Dimitrov et al. 1986) from EH patients and in SHR vascular smooth muscles (Heagerty et al. 1986).

To explain the logic of the next stage of our study, it should be mentioned that at least some of the key chains of the phosphoinositol signaling pathway in the cells are represented by proteins which are products of certain proto-oncogenes of the genome. These oncogene products are localized in the plasma membrane, particularly presented in the form of enzymes, monitoring signal transductions, skeleton shrinking, and the rate of ion carrier translocation (for instance Na^+/H^+ exchange). It may be suggested, therefore, that some abnormally operating proto-oncogenes by themselves or some other loci inherited in conjunction with certain proto-oncogenes of the cell genome could, in principle, explain membrane abnormalities found in primary hypertension (or at least a portion of them; Postnov 1987). Thus, conjecturing the involvement of cellular oncogenes in the pathogenesis of membrane alterations, our group undertook a study of the restriction fragment length polymorphism (RFLP) of DNA on certain proto-oncogene loci, comparing spontaneously hypertensive (SHR) and Wistar-Kyoto normotensive rats (WKY). Analysis of SHR and WKY genomic DNA revealed polymorphism in the c-*src* locus (Kotelevtsev et al. 1988). Next we undertook a study to cosegregate c-*src* alleles with membrane abnormalities in F_2 generation of SHR × WKY) hybrids, using RFLP results as Mendelian traits in linkage studies (Kotelevtsev et al. 1989). As indicators of ion-transport abnormalities Na, K cotransport and Ca content in erythrocytes were used (Kotelevtsev et al. 1987). These parameters, consisting of "intermediate phenotypes," were selected due to their positive correlation with BP (the use of BP itself as a variable in the linkage studies is too complicated because of both the polygenic nature of this trait and the epistatic interactions of genes at individual loci; (Schlager and Chang-Shin 1989).

(SHR × WKY)F_2 hybrids were divided into three groups according to their c-*src* genotypes, as determined by Southern blotting: homozygous (SS; SHR genotype), WW. (WKY genotype) and heterozygous (WS). Genotypes were determined in restriction fragments of DNA (*Hin*dIII and *Pst*I restrictases were used) hybridized with v-*src*.

As can be seen from Fig. 1 the rate of Na^+, K^+ cotransport in the homozygous WW group was equal to that of WKY rats and differed from the heterozygous WS group (which has an intermediate value) and from the homogyzous SS group, in which the rate of cotransport was equal to that of SHR. Thus, a tight linkage between RFLP in c-*src* and the $Na^+ - K^+$ cotransport determining locus was demonstrated.

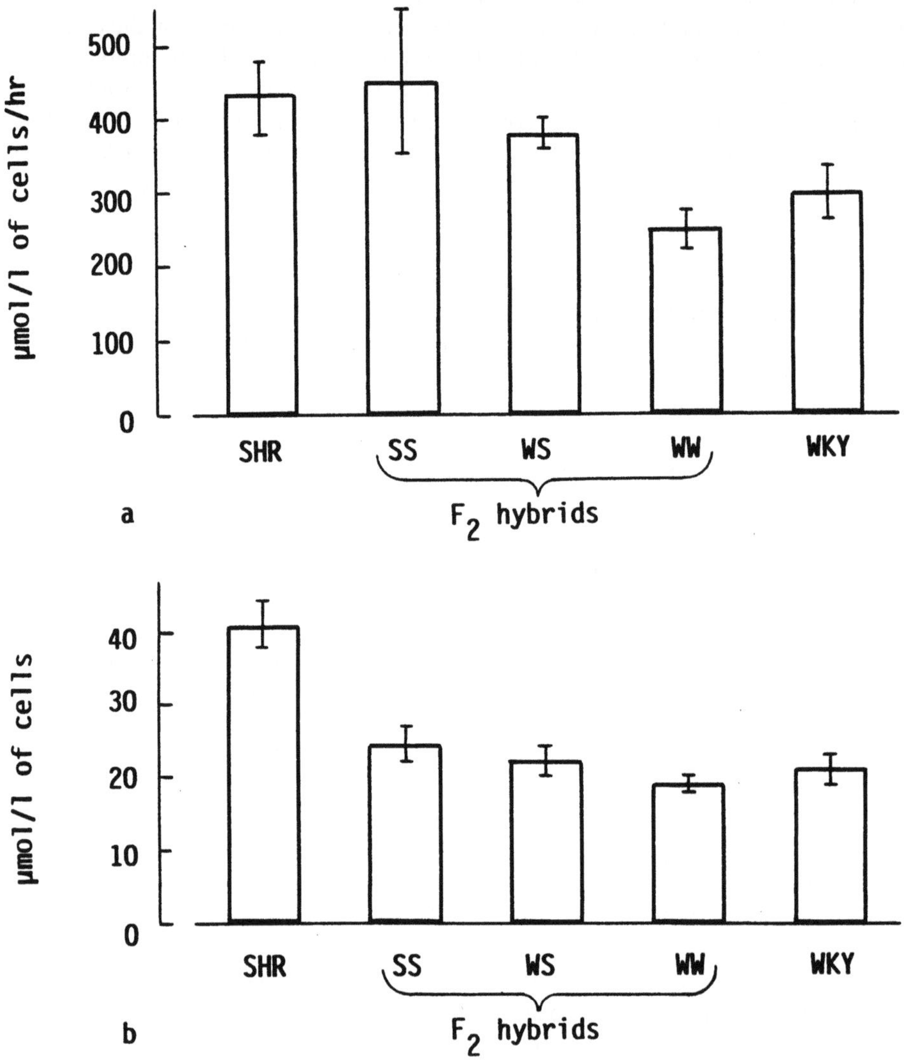

Fig. 1a, b. Na^+, K^+ cotransport (**a**) and calcium content (**b**) in erythrocytes of (SHR × WKY)F_2 hybrids according to c-*src* genotype. F_2 hybrids: *SS*, *WW*, homozygous; *WS*, heterozygous

Calcium accumulation was inherited in a different manner. In the erythrocytes of WW hybrids, calcium content was equal to that of WKY rats and lower than that in the WS and SS groups, which in turn had significantly lower levels than SHR (Fig. 1). This trait is probably under polygenic control, but the c-*src* locus has definite influence upon it. All the results of these studies lead to the conclusion that the locus of DNA marked by the RFLP in c-*src* contains gene(s) which determine the increased $Na^+ - K^+$ cotransport and Ca^{2+} accumulation in the erythrocytes of rats with spontaneous hypertension. These data point in principle to the genomic source of membrane alterations in the case of an animal model of primary hypertension as well as to the involvement of proto-oncogenes in their development.

A comparison of membrane alterations in both forms of hypertension discussed here shows their deep pathogenetic similarity, although it does not mean we view them as identical. At our present stage of knowledge we can suggest that this similarity may reflect the presence in human EH of an initial genomic source of cell membrane abnormalities (at least a portion of them) which constitutes the basis for a predisposition to hypertension. The actual appearance of hypertension in humans (unlike that in SHR, where it is entirely genetically determined) is affected by environmental factors as well.

References

Berk BC, Vallega G, Muslin AJ, Gordon HM, Canessa M, Alexander RW (1989) Spontaneously hypertensive rat vascular smooth muscle cells in culture exhibit growth Na^+/H^+ exchange. Clin Invest 83: 822–829

Berridge MJ (1984) Inositol triphosphate and diacrylglycerol as a second messenger. Biochem J 220: 345–360

Boriskina GM, Gulak PV, Postnov YV (1978) Phosphoinositide in the erythrocyte membrane of rats with spontaneous and renal hypertension. Experientia 34: 744

Canessa M, Adragna N, Solomon HS, Connoly TM, Tosteson DC (1980) Increased sodium-lithium countertransport in red cells of patients with essential hypertension. N Engl J Med 302: 772–776

Dimitrov D, Resink TJ, Müller FB, Bühler FR (1986) Altered platelet phosphatidyl inositol metabolism in essential hypertension. J Hypertens 4 [Suppl 6]: S346–S348

Garay RP, Meyer PA (1979) A new test showing abnormal net Na^+ and K^+ fluxes in erythrocytes of essential hypertensive patients. Lancet i: 349–353

Guyton AC (1980) Circulatory physiology. III. Arterial pressure and hypertension. Saunders, Philadelphia

Heagerty AM, Ollerenshaw JD, Swales JD (1986) Abnormal vascular phosphoinositide hydrolysis in the spontaneously hypertensive rat. Br J Pharmacol 89: 803–807

Jones AW (1973) Altered ion transport in vascular smooth muscle from spontaneously hypertensive rats. Influence of aldosterone, norepinephrine and angiotensin. Circ Res 33: 563–572

Kotelevtsev YV, Orlov SN, Pokudin NI, Agnayev VM, Postnov YV (1987) Genetic analysis of Na^+, K^+-cotransport calcium content and blood pressure in (SHR × WKY)F_2 hybrids. Bull Exp Biol Med 103: 456–458 (in Russian)

Kotelevtsev YV, Brashishkite DA, Spitkovsky DD, Kiselev FL, Postnov YV (1988) Interstrain restriction fragment length polymorphism of c-fos and c-src oncogene loci in spontaneously hypertensive and normotensive rats. J Hypertens 6: 781–799

Kotelevtsev YV, Spitkovsky DD, Orlov SN, Postnov YV (1989) Interstrain restriction fragment length polymorphism in the c-src correlates with Na, K-cotransport and calcium content in hybrid rat erythrocytes. J Hypertens 7 [Suppl 6]: S112–S113

Lang GF (1950) Hypertension. Medgiz, Moscow (in Russian)

Livne A, Veitch R, Grinstein S, Balfe JW, Marquez-Julio A, Rothstein A (1987) Increased platelet $Na^{+} - H^{+}$ exchange rates in essential hypertension: application of a novel test. Lancet i: 533–536

Marche P, Kutouzov S, Girard A, Elghozi JL, Meyer P, Ben-Ishay (1985) Phosphoinositide turnover in erythrocyte membranes in human and experimental hypertension. J Hypertens 3: 25–30

Orlov SN, Postnov IY, Pokudin NI, Kukharenko VY, Postnov YV (1989) $Na^{+} - H^{+}$ exchange and other ion-transport systems in erythrocytes of essential hypertensives and spontaneously hypertensive rats: a competative analysis. J Hypertens 7: 781–788

Pickering GW (1977) Personal views on mechanisms of hypertension. In: Genest J, Koiw E, Kuchel O (eds) Hypertension. Physiopathology and treatment. McGraw-Hill, New York, pp 598–606

Postnov YV (1987) Membrane defect in primary hypertension: a reflection of altered cellular oncogene function? Kardiologija 27(11): 98–102 (in Russian)

Postnov YV, Orlov SN (1979) Alteration of cell membranes in hypertension: role of altered membrane control over intracellular calcium. In: First joint US-USSR symposium of hypertension. National Institute of Healthy, pp 182–193

Postnov YV, Orlov SN (1984) Cell membrane alteration as a source of primary hypertension. J Hypertens 2: 1–6

Postnov YV, Orlov SN (1985) Ion transport across plasma membrane in primary hypertension. Physiol Rev 65: 904–945

Postnov YV, Orlov SN, Gulak PV, Shevchenko AS (1976) Altered permeability of the erythrocyte membrane for sodium and potassium in spontaneously hypertensive rats. Pflügers Arch 365: 257–263

Postnov YV, Orlov SN, Shevchenko AS, Adler AM (1977) Altered sodium permeability, calcium binding and Na-K-ATPase activity in the red blood cell membrane in essential hypertension. Pflügers Arch 371: 263–269

Postnov YV, Orlov SN, Pokudin NI (1981) Alteration of the intracellular calcium pool of adipose tissue in spontaneously hypertensive rats. No effect of peripheral immunosympathectomy. Pflügers Arch 390: 256–259

Postnov YV, Kravtsov GM, Orlov SN, Kotelevtsev YV, Pokudin NI, Postnov IY (1987) Regulation by protein kinase C of erythrocyte shape, volume and passive, sodium permeability: alteration in essential hypertension. J Hypertens 5 [Suppl 5]: S257–S259

Postnov YV, Kravtsov GM, Orlov SN, Pokudin NI, Postnov IY, Kotelevtsev YV (1988) Effect of protein kinase C activation on cytoskeleton and cation transport in human erythrocytes. Reproduction of some membrane abnormalities revealed in essential hypertension. Hypertension 12: 267–273

Shlager G, Chang-Shin (1989) The role of dominance and epistasis in the genetic control of blood pressure in rodent models of hypertension. In: 6th International Symposium on SHR and related studies, Iowa City, p 32

Molecular Genetics and Familial Arterial Hypertension

F. Soubrier, X. Jeunemaître, B. Rigat, F. Cambien, and *P. Corvol*

Summary

Human essential hypertension is a heterogeneous disorder of multifactorial origin. Genetic susceptibility plays an important role in the development of the disease and it has been estimated, from family and twin studies, that about 30 % of the observed variation in blood pressure in the population is genetically determined. The unimodal distribution of blood pressure in the general population as well as in the offspring of hypertensive parents, suggests that several genes are probably involved in genetic predisposition. Polymorphic variations in the structure of the genes involved in blood pressure regulation may be the origin of hypertension. One way of identifying these genes is to examining the role of potential candidate genes which influence blood pressure regulation and for which molecular molecular probes are available.

We have summarized in this chapter the results of three different genetic studies in which we have evaluated the effect of allelic variation on candidate genes, either on blood pressure level or on the phenotypic expression of these genes.

Introduction

The molecular mechanisms involved in blood pressure regulation appear each day to be more complex. In addition to the classical renin angiotensin and nervous system, more recent concepts have emerged about the role of hormones or mediators involved in vascular smooth muscle tone regulation and hydroelectrolytic homeostasis (ANF, endothelin, EDRFs, etc.). It appears quite difficult to distinguish the effect of one factor and to assess its responsibility in the development of essential hypertension. However, in spite of the complexity of blood pressure regulation mechanisms, we have to determine which genes are involved in genetic predisposition to blood pressure elevation and by which molecular mechanisms their allelic variations are able to cause this elevation. Since the phenotypic expression of known genes is, in most cases, greatly modified by environmental factors and is influenced by other systems, it is tempting to look for a link between genetic variation (at the genomic level), as analysed by polymorphism identification, and variation in blood pressure.

K. Berg, V. Bulyzhenkov, Y. Christen, P. Corvol (Eds.)
Genetic Approaches to Coronary Heart Disease
and Hypertension

Different approaches can be used to find a relationship between a gene and blood pressure level. But all these approaches are confronted with the same difficulties. Blood pressure is a highly variable parameter, even when measured in standardized conditions. For this reason it would be important to find more accurate and consistent measurements which might allow a better evaluation of blood pressure level. Indeed, recents studies using 24-h ambulatory blood pressure recordings have shown that systolic blood pressure was higher in normotensive subjects having two hypertensive parents than in normotensive subjects having two normotensive parents, by contrast, pressor responses to laboratory stressors were similar in the two groups (Ravogli et al. 1990). These results show the value of ambulatory measurement and indicate that genetic predisposition to hypertension modifies the baseline adjustment of blood pressure more than reactivity to stress. Furthermore, echographic measurements of the left ventricular mass index were correlated with ambulatory blood pressure, addressing the question of the origin of the increase of the ventricular mass, which seems to antedate hypertension development. Longitudinal studies have also shown that environmental factors will alter, more and more through life, the blood pressure level that could be expected from the genetic background (Perusse et al. 1989). Two kinds of confounding factors may influence the short- or long-term blood pressure level and make genetic studies difficult to interpret. The first factors are environmental influences such as alcohol or sodium intake; the second factors are other pathologies, possibly due to other genes, which induce hypertension at various stages of their evolution, such as diabetes or obesity. It is obvious that the search for contributory loci in hypertension will be complicated if these confounding factors are not controlled by careful patient selection.

Different strategies are possible to identify genes involved in blood pressure elevation. Until now nearly all results published on molecular genetics of hypertension concern the involvement of candidate genes. We review these different works below, as well as the possibility of evaluating allelic effects on phenotype expression of candidate genes and their relation with blood pressure. Finally, we examine the conditions required to screen the whole genome for discovering quantitative loci for high blood pressure.

Candidate Genes and Hypertension

The renin gene has been the most studied in hypertension because it is the limiting enzyme of the biochemical reactions leading to the formation of the potent vasoactive angiotensin II, because its expression is finely tuned and because it has been shown that an increase of renin expression is able to provoke hypertension. Several restriction fragment length polymorphisms (RFLPs) have been described, constituting gene markers (Frossard et al. 1986a, b). Morris and Griffiths (1988) were the first to compare the frequency of renin gene polymorphisms in hypertensive and normotensive. They used retrospective data on anti-hypertensive drugs intake, collected in a questionnaire distributed to blood donors, to distinguish the hypertensive population which was compared to the whole cohort of blood donors. They could not detect any differences in the allele frequency of the *Hin*dIII polymorphism they studied between the two groups.

More recently, Naftilan et al. (1989) reported the absence of association between the renin gene and hypertension in large pedigrees of Utah. Two types of analysis were performed in this work. The first was a study of the cosegregation of renin gene haplotypes and high blood pressure in large pedigrees; the second was to search for an association between haplotypes and blood pressure subgroups in the individuals from large pedigrees. The results of both analyses were negative.

We performed a case-control study on the frequency of renin genotypes and haplotypes in the hypertension clinic of the Broussais Hospital in Paris (Soubrier et al. 1990). We selected two contrasting groups both for their blood pressure and for their family history of blood pressure. The hypertensive group (102 patients) had a diastolic blood pressure above 95 mmHg at two consecutive visits, no cause of secondary hypertension, hypertension before age 65 years in at least one parent and one sibling, and their mean age was 45.3 ± 10 years. The normotensive group (120 subjects) had systolic and diastolic blood pressure under 140 mmHg and 90 mmHg, respectively, and no familial history of blood pressure. All subjects were Caucasiany had a body mass index under 27 kg/m^2, and no diabetes mellitus or renal insufficiency.

Using both a human renin cDNA probe and a genomic fragment of the renin gene, we analyzed three DNA polymorphisms of the renin gene by digesting genomic DNA with *Taq*I, or *Hin*dIII. These three restriction enzyme sites are located in the 5′ part of the gene, in the first intron and in the 3′ part of the gene,

Table 1. Estimated renin gene haplotype frequencies defined by the combination of the three RFLP sites in hypertensives and normotensives. (From Soubrier et al. 1990)

	Hypertensives		Normotensives		All		
Haplo-types[a]	Estimated frequency	Estimated number (n = 204)	Estimated frequency	Estimated number (n = 240)	Estimated frequency	Estimated number (n = 444)	N^b
HIT	0.429	87.5	0.453	108.7	0.443	196.7	218.4
Hit	0.085	17.3	0.035	8.4	0.057	25.3	23.4
HIT	0.177	36.1	0.195	46.8	0.186	82.6	57.1
Hit	0.0	0.0	0.0	0.0	0.0	0.0	6.1
hIT	0.277	56.5	0.245	58.8	0.257	114.1	99.5
hit	0.022	4.5	0.042	10.1	0.036	16.0	10.7
hiT	0.009	1.8	0.020	4.8	0.017	7.5	26.0
hit	0.000	0.0	0.010	2.4	0.004	1.8	2.8
Loglik	222.4		269.8		495.6		

Comparison between normo- and hypertensives: χ^2_6 = 2 (495.6 − 222.4 − 269.4) = 7.6, NS

[a] Capital letters designate the frequent allele. H/h, T/t, and I/i for the *Hin*dIII, *Taq*I, and *Hin*fI polymorphisms, respectively

[b] N is the expected number under the hypothesis of independence between markers

respectively. We compared the genotype frequencies for the three RFLPs and found no differences between the hypertensives and the normotensives. We also estimated, by the maximum likelihood method, the frequency of the possible RFLP combinations into haplotypes. By combining the three RFLPs into haplotypes, the heterozygosity of the renin genetic markers was greatly increased. However, the estimated frequencies of the renin haplotypes in normotensives and hypertensives were not significantly different, although some differences could be detected for rare haplotypes (Table 1).

We used the same renin gene markers in a differently designed study, using the affected sib-pair method described by Lange (1986). We analysed the genotypic resemblance for the renin markers in 98 hypertensive pairs, selected on the same clinical criteria (X. Jeunemaître et al. manuscript in preparation). To use this method it is necessary to know the haplotype frequency in the population studied; we used the frequency that was determined previously in the case control study. The frequencies of total, partial, and no haplotype concordances were calculated for the 98 pairs and compared to the expected concordances, computed under the hypothesis of absence of cosegregation of the markers and the disease. No significant difference between the expected and the calculated concordance for the renin haplotypes was observed, indicating that there is no linkage between the renin gene and hypertension in this study.

The possibility of finding a positive result in such studies depends on three factors:

1. the contribution of the candidate gene to the phenotypic trait,
2. the informativity of the markers and
3. the size of the population studied.

Our negative results indicate that the contribution of the renin locus to hypertension, if any, cannot be detected using these markers and this size and type of population. The possibility remains that a small contribution of the renin gene in essential hypertension might be detected by using a larger population or more informative markers that we will consider later. However, considering hypertension as a homogeneous disease is artificial and we can also imagine that, in a small subset of patients, the renin locus might exert a greater contribution.

Both our results and those Naftilan et al. may appear a bit disappointing when compared to the results obtained by Rapp et al. (1989) on the Dahl rat and by Kurtz et al. (1990) on spontaneously hypertensive rats (SHR). Both groups have analyzed the F2 generation obtained by crossing rats from inbred hypertensive strains (Dahl-sensitive or SHR) with normotensive rats (Dahl resistant or inbred Lewis rat) to obtain F1 rats and then by intercrossing F1 rats to obtain F2 rats. In each study a renin gene allele was identified by a RFLP and was analyzed for cosegregation with blood pressure in the F2 generation. The F2 generation rats were separated into three groups, according to their renin genotypes: homozygotes for the hypertensive strain allele, heterozygotes for the hypertensive gene allele and the normotensive alleles and homozygotes for the normotensive renin gene allele. A clear-cut blood pressure difference was observed by Rapp et al.

(1989) between the three genotype groups in the Dahl F2 rats. A significant difference in blood pressure was also observed between F2 rat heterozygotes for the renin alleles and F2 rat homozygotes for the Lewis normotensive allele by Kurtz et al. (1990). However, the results on the SHR rats were not confirmed by Lindpaintner et al. (1990) who studied inbred SHR stroke-prone rats and their normotensive controls of the Wistar-Kyoto strain. These authors were not able to find any blood-pressure difference between the three renin genotype subgroups; moreover, rats with the hypertensive renin allele were able to better modulate their renin secretion in response to a sodium load. The discrepancy between these results could be explained by the use of different normotensive rat strains which carry different renin alleles.

For a long time the HLA locus has been considered as a candidate region for high blood pressure and has been frequently tested for its association with hypertension. Indeed 20 studies showing associations of specific HLA antigens with hypertension were recorded in the recent work of Gerbase-DeLima et al. (1989). In this same work, the authors have analyzed the concordance for HLA haplotypes inheritance of 98 hypertensive sib-pairs. Patients between 17 and 52 years of age were selected with systolic or diastolic pressure above 160 and 95 mmHg, respectively. Patients were from different ethnic origins: 18 families were Caucasian, 12 were negroid and 1 was of Japanese origin. They were typed for HLA A and B antigens. In this work, the observed frequency of shared HLA haplotypes was different from the number and this difference was highly significant ($p = 0.0009$). These results indicate a distorted segregation of HLA haplotypes in hypertension. However, in 11 of 31 families there was no linkage between HLA and hypertension, and that could be explained by the heterogeneity of hypertension, resulting both from the role of environmental factors and/or of different genes. These interesting results need to be confirmed by other studies, which should examine other populations.

The use of RFLP for genetic analyses of candidate genes might be rapidly replaced by the direct search for mutations either in the coding region or in regulatory regions of these genes. These new developments are made possible by both new methods for detection of mismatch between DNA strands (Cotton et al. 1988; Ganguly and Prockop 1990) and direct sequencing of PCR products (Gyllensten and Erlich 1988).

Due to the great number of genes involved in blood pressure regulation, it is obvious that several other genes or candidate loci have to be tested, and that it will be necessary to develop powerful genetic markers to be able to detect even small contributions of these genes to the phenotypic trait.

Allelic Effects on Biological Markers of Hypertension

One way to implicate a gene in the development of hypertension is to demonstrate a modification of its expression in relation to hypertension development, either quantitatively or qualitatively. In contrast to the approaches described above, intermediate phenotypes are used here instead of genetic markers to evaluate gene expression and its possible link with hypertension. To be used in genetic studies, an intermediate phenotype should fulfill some conditions. First, the phenotype should be stable, without short-term variations, or allow a reliable measurement in standardized conditions. This would allow analysis of long term modulation, such as a genetic effect. Secondly, the phenotype should reflect directly the expression of one gene. In the case of complex phenotypes, involving more than one gene or influenced by several biochemical steps – such as protein processing, secretion, hormones and receptors interactions – analysis of the gene effect would be very difficult. To demonstrate that a phenotype is involved in hypertension, at least two criteria should be satisfied. The first criterion is that a difference in this phenotype is observed between hypertensives and normotensives that could precede the presence of hypertension. Secondly, the phenotype should be directly influenced by identified alleles, which would represent the genetic basis of liability to hypertension. Indeed, the first criterion is not sufficient as phenotypic modifications could be the consequence of hypertension itself and might not be the result of genetic predisposition.

Some phenotypes have been extensively studied and it is interesting to evaluate how they have met the above criteria.

The Sodium-Lithium Countertransport. This phenotype is related to regulation of the Na^+ content of erythrocytes which has been found to be higher in hypertensives (Canessa et al. 1980). Genetic analysis of $Na^+ - Li^+$ countertransport in families with hypertension or myocardial infarction from Utah has shown that a major genetic locus explained 34 % and polygenic inheritance explained 46 %, of the phenotypic variance. Random environmental effects explained the remaining 20 % (Hasstedt et al. 1988). In this study, less than 1 % of blood pressure variance was explained by this locus. A case-control study was performed independently by Turner et al. (1980) in 1475 caucasians. They found that two different normal distributions of $Na^+ - Li^+$ countertransport are present in this population and that an increase of the countertransport makes only a small contribution to the prediction of higher blood pressure. It is likely that more than one gene is involved in this phenotype and one candidate is the Na^+ - H^+ antiporter gene, which has been cloned (Sardet et al. 1989). However, it was reported in a recent paper that there is no relation between alleles at this locus and either the level of the Na^+ - Li^+ countertransport or familial elevation of blood pressure (Lifton et al. 1991).

The Urinary Kallikrein Phenotype. Interest in the urinary kallikrein gene developed as a result of several studies which showed that urinary kallikrein excretion was lower in hypertensives than in normotensives (Zinner et al. 1976). Similar, results were obtained in hypertensive rats, in which lower UK level is observed when they develop hypertension (Zinner et al. 1976). It was also shown that the UK phenotype was under genetic control and that 51 % of phenotype variance was due to a dominant allele present in 28 % of individuals. However, it will be necessary to use genetic markers of the renal kallikrein gene to demonstrate that low kallikrein excretion is a genetically determined phenotype predisposing to hypertension.

The Renin Angiotensin System Genes. Concentration of the gene products in plasma is the only available measurement. The great variability of plasma active renin concentration, modified by short term variations, has hampered studies concerning the long-term regulation of this parameter, including the genetic control. However, plasma renin can be measured under standardized clinical and metabolic conditions (Sealey et al. 1990). For genetic studies of renin phenotype, total plasma renin is a more reliable measurement since it is less concerned by short term variations, such as postural variations (Toffelmire et al. 1989). Plasma angiotensinogen concentration and plasma angiotensin-I converting enzyme (ACE) concentration are stable in a given individual. Plasma angiotensinogen was found to be higher in plasma of hypertensives than in normotensives in the classical work of Walker et al. (1979), but this parameter has not been investigated in family studies.

The genetic control of ACE, on the other hand, has been extensively investigated. A first study was conducted in normal nuclear families. The results demonstrated a familial resemblance for plasma ACE level (Cambien et al.

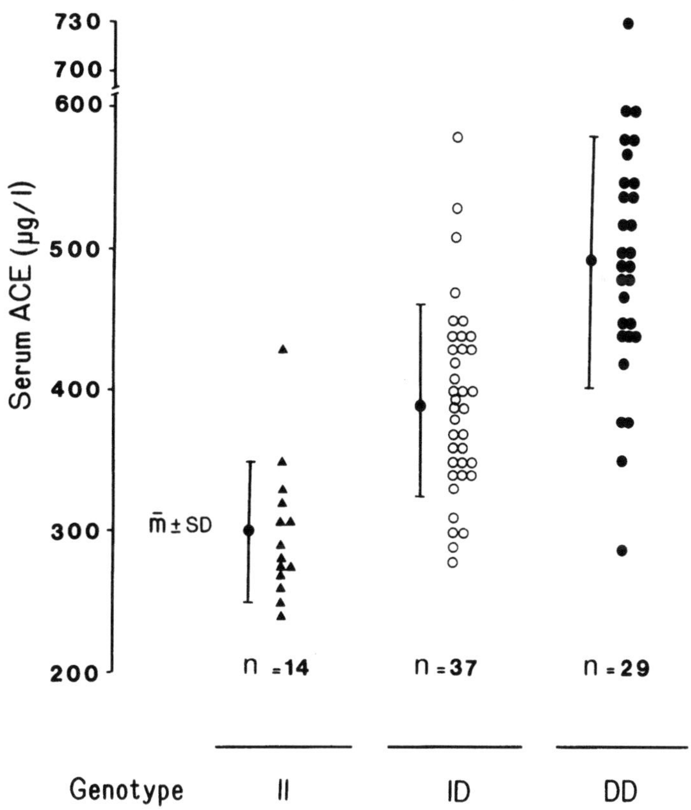

Fig. 1. Serum immunoreactive ACE concentrations (µg/l) for individuals with the II, ID, and DD genotypes shown in the *left, middle,* and *right* panels repectively. *Solid vertical bars* indicate mean concentration and standard deviation for each group. (From Rigat et al. 1990, with permission)

1988). This study showed that this familial resemblance was compatible with the effect of a major gene. Using the human endothelial ACE cDNA that we have cloned (Soubrier et al. 1988), we found a polymorphism of the ACE gene, consisting of the presence or absence of a 250 bp insert, detected by several restriction enzyme digests. We looked for an association between this polymorphism and the level of circulating ACE in 80 healthy subjects (Rigat et al. 1990). In this group, the frequencies of the shorter and the longer allele were 0.6 and 0.4, respectively. Serum ACE level was measured by a direct radioimmunoassay (Alhenc-Gelas et al. 1983). No correlation was found between serum ACE level and any of the clinical parameters tested (gender, systolic and diastolic blood pressure, age, body mass index). A marked difference in serum ACE levels was observed between subjects in each of the three ACE genotype classes. Serum immunoreactive ACE concentrations were, respectively, 299 ± 49, 392 ± 66.8, and 494 ± 88.3 μg/liter for homozygotes with the longer allele (n = 14), heterozygotes (n = 37) and homozygotes with the shorter allele, respectively (n = 29) (Fig. 1). The polymorphism accounted for 47% of total variance of the phenotype. Interindividual differences are thus in a large part explained by this genetic effect, and a more accurate reference interval for circulating ACE concentration can be given for each ACE genotype class.

A more complex phenotype is the response of aldosterone and of intra-renal vessels to angiotensin II infusion during sodium restriction and sodium charge, studied by the group of Hollenberg and Williams (1990). An abnormal response was found in 50% of hypertensives and was the basis for a classification into "modulators" and "nonmodulators." This abnormality of response appeared to be genetically transmitted, was more frequent in hypertensives having a familial history of hypertension than in hypertensives with no such history and was corrected by ACE inhibitors.

The points of interest and the limitations of these studies are apparent. Among the points of interest are the fact that a stable biological phenotype is more reliable than blood pressure measurement and that detection of a phenotype that would precede blood pressure elevation would be invaluable. Among the limitations is the fact that only candidate genes expression can be evaluated, and that does not permit implication of other, and possibly unknown, genes.

Finally, important intermediate phenotypes for hypertension might not be accessible to usual measurements since primary abnormalities might be restricted to some cellular function which has no consequence on any circulating parameter.

The Whole Genome Approach

The search for a locus responsible for a major contribution to blood pressure regulation, out of the known candidate genes, is the challenge of genetic investigation of hypertension for the coming years. Indeed we know that the mapping of loci contributing to quantitative phenotypes is possible and has been performed in plants (Paterson et al. 1988). The problem of blood pressure is more complex but there is no theoretical obstacle to map the blood pressure loci. However this ambitious project will not be possible without improvement of existing techniques and development of new tools. Choices have to be made concerning the patients to be included, the type of genetic study and the markers used.

1. Patients: heterogeneity of the disease will be one of the major problems encountered in such a project and phenotypic classification would be interesting but is not available now. We have tried to select patients in our studies with isolated hypertension to avoid complex interactions with other diseases or environmental factors and it is probably the best way to find a significant contributory locus.
2. Genetic study: only family studies are possible in such a genetic approach, where the genetic markers may be at a long genetic distance from the disease locus. The affected sib-pairs method has been used for other genetic disease, such as diabetes mellitus (Cox et al. 1988). This method allows analysis of small families, where at least two sibs are affected; analysis of nonaffected sibs is not required.
3. Genetic markers: the more polymorphic markers are the variable number of tandem repeats (VNTR) described by Nakamura et al. (1987) but they are preferentially localized in some parts of the chromosome.

For this reason, the discovery of highly polymorphic and widely distributed microsatellites in the genome, composed of short dinucleotide repeats (CA)n or (TG)n, where n is highly variable, opens an opportunity to test the whole genome (Weber and May 1989). However, detection of the polymorphisms requires DNA amplification and electrophoresis of the sample on acrylamide denaturing gels. Even if multiple detections are possible on the same electrophoresis, the number of polymorphisms to be detected in around 600 patients requires the conception of automatic devices to analyze the detection of the polymorphisms.

Conclusions

Phenotype studies are promising if there is evidence that their modifications during hypertension are of genetic origin and cause a biological predisposition to hypertension. To demonstrate this point, studies of concerned phenotypes, linked to accountable marker genotypes, in normal and affected families are required.

While it is compulsory to test candidate genes for their participation in blood pressure elevation, this strategy is restricted per se because it will not permit discovery of new genes. This is the major argument for removing hypertension from the ghetto of the candidate genes and to direct efforts towards the whole genome approach.

References

Alhenc-Gelas F, Weare J, Johnson R, Erdös EG (1983) Measurement of human converting enzyme level by direct radioimmunoassay. J Lab Clin Med 101: 83–96

Cambien F, Alhenc-Gelas F, Herbeth B, Andre JL, Rakotovao R, Gonzales MF, Allegrini J, Bloch C (1988) Familial resemblance of plasma angiotensin-converting enzyme level: the Nancy study. Am J Hum Genet 43: 774–780

Canessa M, Adragna N, Solomon H, Connolly TM, Tosteson BS, Tosteson DC (1980) Increased sodium-lithium countertransport in red cells of patients with essential hypertension. N Engl J Med 302: 772–776

Cotton R, Rodrigues N, Campbell D (1988) Reactivity of cytosine and thymine in single-base-pair mismatches with hydroxylamine and osmium tetroxide and its application to the study of mutations. Proc Natl Acad Sci USA 85: 4397–4401

Cox NJ, Baker L, Spielman RS (1988) Insulin gene sharing in sib-pairs with insulin-dependent diabetes mellitus: no evidence for linkage. Am J Hum Genet 42: 167–172

Frossard PM, Gonzalez PA, Dillan NA, Coleman RT, Atlas SM (1986a) Human renin (REN) gene locus: Bgl II, Rsa I and Taq I RFLPs. Nucleic Acids Res 14: 6778

Frossard PM, Gonzalez PA, Fritz LC, Ponte PA, Fiddes JC, Atlas SM (1986b) Two RFLPs at the human renin (ren) gene locus. Nucleic Acids Res 14: 4380

Ganguly A, Prockop DJ (1990) Detection of single-base mutations by reaction of DNA heteroduplexes with water-soluble carbodiimide followed by primer extension: application to products from the polymerase chain reaction. Nucleic Acid Res 18: 3933–3939

Gerbase-DeLima M, DeLima JJG, Persoli LB, Bernardes Silva H, Marcondes M, Bellotti G (1989) Essential hypertension and histocompatibility antigens. Hypertension 14: 604–609

Gyllensten UB, Erlich HA (1988) Generation of single-stranded DNA by the polymerase chain reaction and its application to direct sequencing of the HLA-DQA locus. Proc Natl Acad Sci USA 85: 7652–7656

Hasstedt SJ, Wu LL, Ash KO, Kuida H, Williams RR (1988) Hypertension and sodium-lithium countertransport in Utah pedigrees: evidence for major locus inheritance. Am J Hum Genet 43: 014–022

Hollenberg NK, Williams GH (1990) Abnormal renal function, sodium volume homeostasis, and renin system behavior in normal-renin essential hypertension. In: Laragh JH, Brenner BM (eds) Hypertension: pathophysiology, diagnosis and management. Raven, New York, pp 1349–1370

Kurtz TW, Simonet L, Kabra PM, Wolfe S, Chan L, Hjelle BL (1990) Cosegregation of the renin allele of the spontaneously hypertensive rat with an increase in blood pressure. J Clin Invest 85: 1328–1332

Lange K (1986) The affected sib-pair method using identity by state relations. Am J Hum Genet 39: 148–150

Lifton RP, Hunt SC, Williams RR, Pouyssegur J, Lalouel JM (1991) Exclusion of the Na^+/H^+ antiporter as a candidate gene in human essential hypertension. Hypertension (in press)

Lindpaintner K, Takahasi S, Ganten D (1990) Structural alterations of the renin gene in stroke-prone spontaneously hypertensive rats: examination of genotype-phenotype correlations. J Hypertens 8: 763–773

Morris BJ, Griffiths LR (1988) Frequency in hypertensives of alleles for a RFLP associated with the renin gene. Biochem Biophys Res Commun 150: 219–224

Naftilan AJ, Williams R, Burt D, Paul M, Pratt RE, Hobart P, Chirgwin J, Dzau V (1989) A lack of genetic linkage of renin gene restriction length polymorphisms with human hypertension. Hypertension 14: 614–618

Nakamura Y, Leppert M, O'Connell P, Wolf R, Hom T, Culver M, Martin C, Fujimoto E, Hoff M, Kumlin E, White R (1987) Variable number of tandem repeat (VNTR) markers for human gene mapping. Science 235: 1616–1622

Paterson AH, Lander ES, Hewitt JD, Peterson S, Lincoln SE, Tanksley SD (1988) Resolution of quantitative traits into Mendelian factors by using a complete linkage map of restriction fragment length polymorphisms. Nature 335: 721–726

Perusse L, Rice T, Bouchard C, Vogler GP, Rao DC (1989) Cardiovascular risk factors in a french-canadian population: resolution of genetic and familial environmental effects on blood pressure by using extensive information on environmental correlates. Am J Hum Genet 45: 240–251

Rapp JH, Wang SM, Dene H (1989) A genetic polymorphism in the renin gene of Dahl rats cosegregate with blood pressure. Science 243: 542–544

Ravogli A, Trazzi S, Villani A, Mutti E, Cuspidi C, Sampieri L, De Ambroggi L, Parti G, Zanchetti A, Mancia G (1990) Early 24-hour blood pressure elevation in normotensive subjects with parental hypertension. Hypertension 16: 491–497

Rigat B, Hubert C, Alhenc-Gelas F, Cambien F, Corvol P, Soubrier F (1990) An insertion/deletion polymorphism in the angiotensin I-converting enzyme gene accounting for half the variance of serum enzyme levels. J Clin Invest 86: 1343–1346

Sardet C, Franchi A, Pouyssegur J (1989) Molecular cloning, primary structure and expression of the human growth factor-activatable Na^+/H^+ antiporter. Cell 56: 271–280

Sealey J, Campbell GA, Preibisz JJ (1990) Hormone assays: renin, aldosterone, peripheral vein, renal vein, and urinary assays. In: Laragh JH, Brenner BM (eds) Hypertension: pathophysiology, diagnosis and management, Raven, New York, pp 1443–1460

Soubrier F, Alhenc-Gelas F, Hubert C, Allegrini J, John M, Tregear G, Corvol P (1988) Two putative active centers in human angiotensin I-converting enzyme revealed by molecular cloning. Proc Natl Acad Sci USA 85: 9386–9390

Soubrier F, Jeunemaitre X, Rigat B, Houot AM, Cambien F, Corvol P (1990) Similar frequencies of renin gene RFLPs in hypertensive and normotensive subjects. Hypertension 16: 712–717

Toffelmire EB, Slater K, Corvol P, Ménard J, Schambelan M (1989) Response of plasma prorenin and active renin to chronic and acute alterations of renin secretion in normal humans. J Clin Invest 83: 679–687

Turner S, Weidman W, Michels V, Reed T, Ormson C, Fuller T, Sing C (1989) Distribution of sodium lithium countertransport and blood pressure in caucasians five to eighty-nine years of age. Hypertension 13: 378–391

Walker WG, Whelton PK, Saito H, Russell RP, Hermann J (1979) Relation between blood pressure and renin, renin substrate, angiotensin II, aldosterone and urinary sodium and potassium in 574 ambulatory subjects. Hypertension 1: 287–291

Weber JL, May P (1989) Abundant class of human DNA polymorphisms which can be typed using polymerase chain reaction. Am J Hum Genet 44: 388–396

Zinner SH, Margolius HS, Rosner B, Keiser HR, Kass EH (1976) Familial aggregation of urinary kallikrein concentration in childhood: relation to blood pressure, race and urinary electrolytes. Am J Epidemiol 104: 124–132

Gene Mapping in Experimental Hypertension

T. W. Kurtz, L. Simonet, V. Křen, and *M. Pravenec*

Summary

The spontaneously hypertensive rat (SHR) is the most widely studied animal model of hypertension and it has been proposed that the identification of genes contributing to increased blood pressure in the SHR might shed light on the pathogenesis of essential hypertension in humans. In F2, backcross, and recombinant inbred populations derived from the SHR and inbred normotensive strains, one can search for chromosome regions that contain genes regulating blood pressure by studying the relationship between blood pressure and the pattern of inheritance of:

1. polymorphic markers for specific genes believed to be related to the pathogenesis of hypertension (so-called "candidate" gene approach), and/or
2. a mapped set of polymorphic markers dispersed throughout the rodent genome (so-called "reverse" genetics approach).

Using the "candidate" gene approach, we have found evidence suggesting that, in the rat, sequence variation within or near the renin gene, the kallikrein gene family, and the RT1 gene complex may have the capacity to affect blood pressure. With respect to the "reverse" genetics approach, we are attempting to develop a rough genetic map that can be used to study the pathogenesis of hypertension in SHR. To date, we have identified over 70 polymorphic markers that distinguish between the SHR and a normotensive Brown-Norway (BN) strain. Analysis of a large set of recombinant inbred strains derived from the SHR and BN progenitors indicates that most of these markers are not tightly linked. These observations suggest that, in the SHR, it should be feasible to employ both the "candidate" gene and "reverse" genetics approaches to search for chromosome regions that contain genes regulating blood pressure.

Introduction

In humans, it is well recognized that a substantial portion of the population variation in blood pressure is genetically determined. In addition, it would appear that at least some forms of essential hypertension (increased blood

K. Berg, V. Bulyzhenkov, Y. Christen, P. Corvol (Eds.)
Genetic Approaches to Coronary Heart Disease
and Hypertension

pressure of unknown etiology) may have a strong genetic component (Rapp 1983a). However, the quantitative trait loci (QTLs) which mediate the increased vascular resistance characterizing essential hypertension remain to be identified. In this article, we discuss some of the problems associated with genetic studies of hypertension in humans and review our progress with gene mapping in the spontaneously hypertensive rat.

Gene Mapping in Human Hypertension

A number of formidable obstacles confront investigators attempting to identify the genetic determinants of human essential hypertension. Without careful consideration of these problems, investigators studying hypertension may encounter the same controversies that have plagued attempts to identify genes determining of other complex conditions such as alcoholism or mental illness (Barnes 1989):

1. Essential hypertension is defined on the basis of an arbitrarily established "normal" range for blood pressure and as such does not constitute a specific disease. Thus, in genetic linkage studies, attempts to phenotypically classify subjects with respect to the "disease" of essential hypertension may not be valid.
2. To avoid problems associated with the arbitrary definition of hypertension, attempts have been made to analyze the inheritance of blood pressure levels per se. However, blood pressure is a highly variable trait and the optimal method for determining the blood pressure "level" of a given individual is unclear. The standard approach has been to average just a few clinic measurements of blood pressure; however, such measurements may be of marginal reproducibility and may not accurately reflect a subject's blood pressure throughout the day (Weber 1988). Furthermore, increased blood pressure often does not become manifest until 50–60 years of age and it may be necessary to measure blood pressure in older subjects. In family studies, however, it will be difficult to obtain accurate blood pressure data from older individuals across multiple generations.
3. Hypertension may have multiple causes and it is difficult to control for the numerous environmental and genetic factors that might affect blood pressure. Therefore, in linkage studies, attempts to combine data from different families may not be appropriate. In fact, it is conceivable that multiple forms of hypertension might exist even within a single pedigree. That is, the environmental and genetic factors causing hypertension in one family member may differ from those causing hypertension in another family member. If so, linkage studies designed to investigate the inheritance of increased blood pressure will be extremely problematic.
4. The modes of inheritance of increased blood pressure are unclear.

Gene Mapping in the Spontaneously Hypertensive Rat

Although comparisons between animals and humans should be regarded with caution, it has been proposed that the identification of genes contributing to increased blood pressure in animals might shed light on the pathogenesis of essential hypertension in humans (Lander and Botstein 1989). Fortunately, many of the problems associated with genetic studies of human hypertension are not encountered in studies of experimental animal models of hypertension.

The spontaneously hypertensive rat (SHR), originally derived from a colony of Wistar rats in Kyoto, Japan, is the most widely studied experimental model of hypertension (Okamoto and Aoki 1963). This strain exhibits many of the neural, renal, and biochemical features which characterize some forms of essential hypertension (Frohlich and Trippodo 1981; Guidi and Hollenberg 1987; Williams et al. 1982). Furthermore, studies from our laboratory and from others indicate that the SHR provides a number of advantages for genetic studies of increased blood pressure (Okamoto and Aoki 1963; Kurtz et al. 1990a, b; Pravenec et al. 1989; Tanase et al. 1970; Tanase 1979):

1. it is highly inbred and therefore constitutes a single genetic form of hypertension,
2. it is characterized by a very high degree of genetic determination of hypertension,
3. only a few major loci may largely determine the increased blood pressure,
4. one can control the breeding patterns as well as many of the environmental factors known to affect blood pressure,
5. the onset of hypertension predictably occurs within a short period of time and does not depend on the induction of extreme environmental circumstances, and
6. with appropriate techniques (computerized signal averaging of blood pressure in the unanesthetized, unrestrained state), one can obtain accurate measurements of blood pressure.

Over the past three decades, many investigators have attempted to identify the genetic determinants of spontaneous hypertension by systemically comparing the SHR to its so-called normotensive "control," the Wistar-Kyoto (WKY) rat. As emphasized by Rapp (1987) however, such interstrain comparisons are of little value when attempting to identify the pathogenetic determinants of hypertension. SHR and WKY rats differ with respect to multiple genes, not just those affecting blood pressure. In fact, most of the physiologic and biochemical differences observed between SHR and WKY rats are probably not causally related to their differences in blood pressure (Kurtz and Morris 1987). Many such biochemical and physiologic differences can be explained by genetic drift and the random fixation of alleles at loci that are not involved in the regulation of blood pressure. Furthermore, it is possible that some of the physiologic and biochemical differences observed between SHR and WKY rats could be secondary to their differences in blood pressure. Attempts to circumvent this problem by studying "pre-hypertensive" SHR may not be valid because the blood pressure of SHR appears to be greater than that of WKY rats even at birth (and perhaps in utero; Gray 1984).

Experimental Approaches to the Identification of Genes Regulating Blood Pressure

Approximately 20 years after Okamoto and Aoki reported the development of the SHR, Rapp (1983b) published a paradigm for the detection of primary genetic causes of spontaneous hypertension. Rapp proposed that in segregating populations derived from crosses of an inbred hypertensive strain and an inbred normotensive strain, one could search for chromosome regions that contain genes regulating blood pressure by studying the relationship between blood pressure and the pattern of inheritance of various physiologic or biochemical genetic markers. If the blood pressure of progeny that inherited a certain genetic marker from the hypertensive progenitor strain was greater than that of progeny that inherited the corresponding genetic marker from the normotensive progenitor strain, one could strongly suspect the presence of a blood pressure regulatory gene in the same chromosome region that controlled the physiologic or biochemical genetic marker. With the advent of restriction fragment length polymorphisms (RFLPs), it subsequently became possible to perform these experiments with molecular genetic markers in addition to physiologic and biochemical genetic markers. A limitation of this approach is that it requires the investigator to make a priori assumptions about which genes are most likely to be involved in the pathogenesis of hypertension. However, blood pressure is an extremely complex quantitative trait and, as a consequence, literally hundreds of so-called "candidate" genes for hypertension can be imagined.

Recently, Lander and Botstein (1989) have proposed that in segregating F2 or backcross populations derived from the SHR and inbred normotensive strains, it may be worthwhile to study the relationship between blood pressure and the pattern of inheritance of multiple polymorphic markers spread throughout the genome. This "reverse" genetics approach allows one to systematically search for chromosome segments that contain genes regulating blood pressure without having to make any a priori assumptions as to which loci might be important in the pathogenesis of hypertension. Although the availability of computer programs for linkage mapping and QTL analysis has greatly simplified this approach, a number of problems must be solved before this technique can be applied to the SHR. In the remainder of this review, we summarize our recent work on the "candidate" gene and "reverse" genetics approaches to gene mapping in the SHR and discuss some of the limitations of both of these approaches.

"Candidate" Gene Approach

Renin Studies in F2 Populations

In an F2 population derived from Dahl salt-sensitive (S) and Dahl salt-resistant (R) rats, Rapp and coworkers (1989) found that the blood pressure of rats that inherited an RFLP marking the renin allele of the S strain was greater than that of rats that inherited the corresponding RFLP marking the renin allele of the R strain. This finding suggests that in the rat, sequence variation in the renin gene can affect blood pressure and may contribute to the pathogenesis of hypertension. Of course, it is also possible that the renin RFLP simply serves as a marker for a blood pressure regulatory locus that is linked to the renin gene. However, given that the renin-angiotensin-aldosterone system is an important determinant of sodium balance and vascular resistance, it is reasonable to suspect that the renin gene itself is the blood pressure regulatory gene marked by the RFLP.

The SHR exhibits alterations in the renin-angiotensin-aldosterone system that are similar to those that characterize patients with "non-modulating" hypertension, a common and highly heritable form of essential hypertension (Williams et al. 1982; Guidi and Hollenberg 1987). Based on this information and the findings of Rapp et al. (1989) in the Dahl rat, we sought to determine whether the inheritance of an RFLP marking the renin gene of the SHR was associated with greater blood pressure than inheritance of an RFLP marking the renin gene of a normotensive strain, the inbred Lewis rat (Kurtz et al. 1990a). When setting up the classical crosses, we deliberately used Lewis rats instead of WKY rats because:

1. at the time of breeding, commercially available rats designated "WKY" did not appear to constitute a single inbred strain (Kurtz and Morris 1987; Kurtz et al. 1989); and
2. the possibility of detecting polymorphisms between SHR and Lewis rats appeared to be greater than between SHR and WKY rats (SHR rats being more distantly related to Lewis rats than to WKY rats).

In an F2 population derived from inbred spontaneously hypertensive rats an inbred normotensive Lewis rats, we found the blood pressure to be greater in rats which inherited the SHR renin allele than in rats which inherited only the Lewis renin allele (ANOVA, $p < .01$; Kurtz et al. 1990a). These findings suggest that the renin allele of the SHR strain has a greater capacity to increase blood pressure than that of the Lewis strain.

Renin Studies in Recombinant Inbred Strains

The genetic studies in the SHR and in the Dahl rat demonstrate the feasibility of using F2 populations to test for associations between specific candidate gene polymorphisms and the transmission of increased blood pressure. An alternative approach to the use of segregating F2 populations involves the use of recombinant inbred (RI) strains derived from hypertensive and normotensive progenitor strains. Recombinant inbred strains are produced by brother × sister matings between members of an F2 population that have been derived from crosses of two highly inbred progenitor strains. In RI strains, segregated alleles from the progenitor strains are fixed by the process of inbreeding and, therefore, RI strains can be considered to represent a reproducible F2 generation. One can search for chromosome regions that contain genes regulating blood pressure by determining whether the blood pressure of the RI strains that inherit a certain genotype of the hypertensive progenitor strain is greater than that of the RI strains the inherit the corresponding genotype of the normotensive progenitor strain. A major advantage of recombinant inbred strains is that the blood pressure value of each strain can be determined by averaging the blood pressure measurements of multiple animals in each strain.

In 1982, Pravenec and Kren began to develop RI strains for use in the genetic analysis of spontaneous hypertension (Pravenec et al. 1989). An SHR strain (SHR/Ola, hereafter referred to as SHR; beyond F50) was used as the hypertensive progenitor and an inbred Brown-Norway strain (BN.1×/Cub, hereafter referred to as BN; beyond F20) was used as the normotensive progenitor. The BN strain was chosen because it was not of Wistar origin and, therefore, a large number of polymorphisms were expected to be found in this system. At present, there are two sets of RI strains: the HXB set, derived from SHR females and BN males (24 strains still bred, all beyond F25), and the BXH set, derived from reciprocal crosses (11 strains, also beyond F25). Blood pressure values of these two sets are not different, suggesting no effects of either the Y chromosome or of mitochondrial inheritance on blood pressure.

On Southern blots of *Bgl*II-digested DNA probed for a renin RFLP, we found that the SHR progenitor strain carries a 1.7 kb restriction fragment, whereas the BN progenitor strain carries a 2.8 kb fragment (Fig. 1). We found the median blood pressure of the RI strains that inherited the renin RFLP of the SHR (135

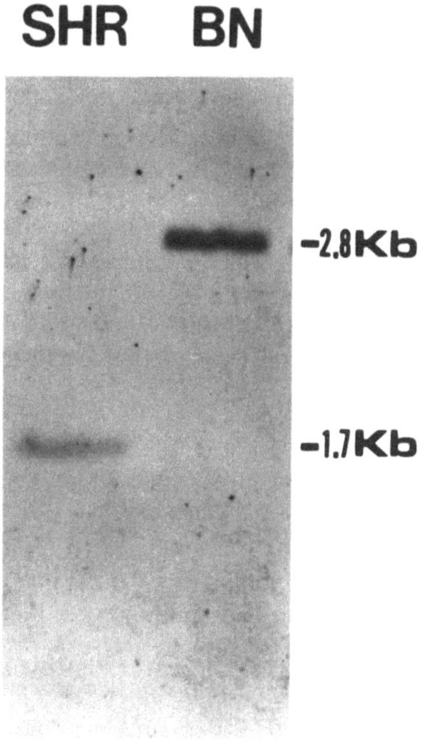

Fig. 1. Southern blot analysis depicting the RFLP used to distinguish the renin alleles of the spontaneously hypertensive rat (*SHR*) and Brown-Norway (*BN*) progenitor strains. The RFLP was detected by probing *Bgl*II digested genomic DNA with a nonisotopically labeled oligonucleotide directed at a variable number of tandem repeats sequence in the first intron of the rat renin gene

mmHg) to be greater than that of the RI strains that inherited the renin RFLP of the BN rat (124 mmHg, $p < .05$; mean arterial pressures; Pravenec et al. 1991a). These studies, like those conducted in segregating F2 populations, provide support for the concept that in the rat, sequence variation in the renin gene, or in closely linked genes, may have the capacity to affect blood pressure. Recently, Lindpainter et al. (1990) found the blood pressure of F2 rats that inherited the renin allele of a stroke prone SHR strain to be similar to that of F2 rats that inherited the renin allele of a European strain of WKY rats. Taken together, the studies in the SHR along with those in the Dahl strain would appear to indicate that, in the rat, a variety of different renin alleles exists and that some of these alleles (but not all) may have different capacities to affect blood pressure. It should also be recognized that the genetic backgrounds of the different normotensive strains used to establish the segregating populations may influence the capacity of a given renin allele to affect blood pressure.

In our studies in the SHR and in the study by Rapp et al. in the Dahl rat, only a portion of the difference in blood pressure between the hypertensive progenitor strains and the normotensive progenitor strains could be attributed to sequence variation at or near the renin locus. Given that multiple genes may be involved in the pathogenesis of spontaneous hypertension (Rapp 1983a), sequence variation at a single locus would not be expected to account for all of the difference in blood pressure between two genetically divergent strains. In the study in Dahl rats (Rapp et al. 1989), the capacity of the renin gene polymorphism to affect blood pressure appeared to be greater than that observed in the studies in SHR (Kurtz et al. 1990a; Pravenec et al. 1991a).

It is difficult to draw conclusions regarding the extent to which sequence variation in the rat renin gene might contribute to the pathogenesis of high blood pressure. Under some genetic and environmental circumstances, a given allele might exert a large effect on blood pressure whereas under other genetic and environmental circumstances, the same allele might appear to have little or no effect on blood pressure. In the studies in the SHR in which the renin gene polymorphisms appeared to exert rather modest effects on blood pressure (<10 mmHg), the diet content of NaCl ranged between 0.5% and 1%. It is conceivable that under the circumstance of an increased dietary intake of NaCl (e.g., with an 8% NaCl diet similar to that employed in studies of the Dahl rat), the same renin gene polymorphisms might exert greater effects on blood pressure.

Renin Gene Polymorphisms and Blood Pressure Regulation – Potential Mechanisms

The mechanisms whereby different renin alleles might have different effects on blood pressure are unknown. Although the first intron of the rat renin gene is characterized by the presence of tandemly repeated copies of a short sequence of DNA that gives rise to extensive restriction fragment length polymorphism, there is little if any evidence to suggest that this variable number of tandem repeats (VNTR) sequence is of particular functional significance (Samani et al.

1989). In hypertensive and normotensive strains of rats, it remains to be determined whether important sequence variation exists within the coding region for renin or within regulatory domains that flank the coding region. The recent finding of low plasma renin activity in transgenic rats made hypertensive by insertion of the mouse Ren-2 gene suggests that altered tissue-specific expression of renin may play a role in the pathogenesis of hypertension (Mullins et al. 1990).

The Rat Renin Gene Chromosome Localization

In the genetic studies conducted to date, it has been assumed that the renin gene itself is the blood pressure regulatory locus being detected by the renin RFLP. However, it is also possible that the renin RFLP is simply marking a blood pressure regulatory gene linked to the renin locus. If so, information regarding the chromosomal location of the rat renin gene could prove useful for identifying other candidate genes in hypertension. Accordingly, we recently used a panel of rat × mouse somatic cell hybrids to map the chromosomal location of the rat renin gene (Pravenec et al. 1991a). We found that, in the rat, the renin gene is located on chromosome 13 and that it belongs to a conserved synteny group located on chromosome 1 in man and mouse. Analysis of the recombinant inbred strains revealed close linkage between the renin gene and the fumarate hydratase gene on linkage group X and, therefore, other genes of linkage group X could also be assigned to chromosome 13 (Pravenec et al. 1991a).

Other "Candidate" Genes

Angiotensinogen

Angiotensinogen, the only known substrate of renin, is present in a variety of tissues including the liver, brain, kidney, and blood vessels. The recent report of hypertension in transgenic mice that carry both the rat renin and angiotensinogen genes together (but not separately) suggests a possible role for the angiotensinogen gene in the pathogenesis of hypertension (Ohkubo et al. 1990). Accordingly, in the RI strains derived from the SHR and BN rats, we sought to determine whether an RFLP in the rat angiotensinogen gene cosegregated with hypertension. The rat angiotensinogen gene is located on chromosome 19 (Mori et al. 1989) and contains a PvuII RFLP that distinguishes between the SHR and BN strains (Fig. 2). However, analysis of the blood pressure data of the RI strains classified with this RFLP failed to reveal an effect of angiotensinogen genotype on blood pressure; the median blood pressure of the RI strains that inherited the SHR angiotensinogen gene (129.4 mmHg) was similar to that of the RI strains that inherited the BN angiotensinogen gene (131.5 mmHg). It is possible, however, that under different genetic or environmental circumstances, this particular polymorphism, or other polymorphisms in the angiotensinogen gene, may have the capacity to affect blood pressure.

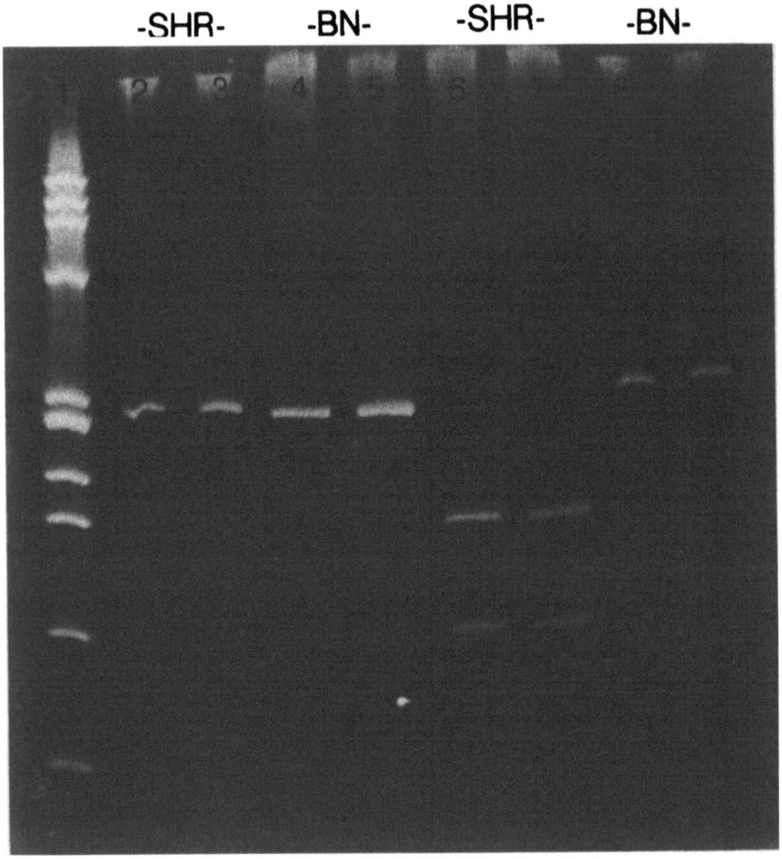

Fig. 2. Polymerase chain reaction (PCR) analysis of a *Pvu*II restriction site in exon 2 of the rat angiotensinogen gene. *Lanes 2* and *3* contain undigested PCR product from two SHR. *Lanes 4* and *5* contain undigested PCR product from two BN rats. *Lanes 6* and *7* contain *Pvu*II-digested product from two SHR. *Lanes 8* and *9* contain *Pvu*II-digested product from two BN rats

Kallikrein Gene Family

A number of investigators have suggested that sequence variation in the gene coding for tissue kallikrein might be involved in the pathogenesis of hypertension (Woodley-Miller et al. 1989). However, molecular evidence of an association between a sequence alteration in the kallikrein gene family and the transmission of increased blood pressure has never been reported. In a recent study using the recombinant inbred strains derived from the SHR and BN strains, we have found evidence suggesting that polymorphisms in the kallikrein gene family may cosegregate with blood pressure (Pravenec et al. 1991b). Figure 3 depicts an RFLP in the rat kallikrein gene family detected by probing NdeI digested genomic DNA with a cDNA probe encoding the 3′ end of the messenger RNA for rat pancreatic kallikrein. In the RI strains that inherited the kallikrein RFLP from the SHR progenitor strain, the median blood pressure (137.8 mmHg) was greater than in the RI strains that inherited the kallikrein RFLP from the BN progenitor strain (123.9 mmHg, $p < .01$; mean arterial arterial pressures). It should be recognized, however, that this preliminary finding could represent a false positive result and, therefore, it will be necessary to test other segregating populations before any firm conclusions are drawn about the role of this chromosome region in the regulation of blood pressure (see section on statistical considerations).

Genes Associated With the Major Histocompatibility Complex

In a recent study in 96 hypertensive siblings distributed among 31 families, Gerbase-DeLima et al. (1989) found a distorted segregation pattern of human

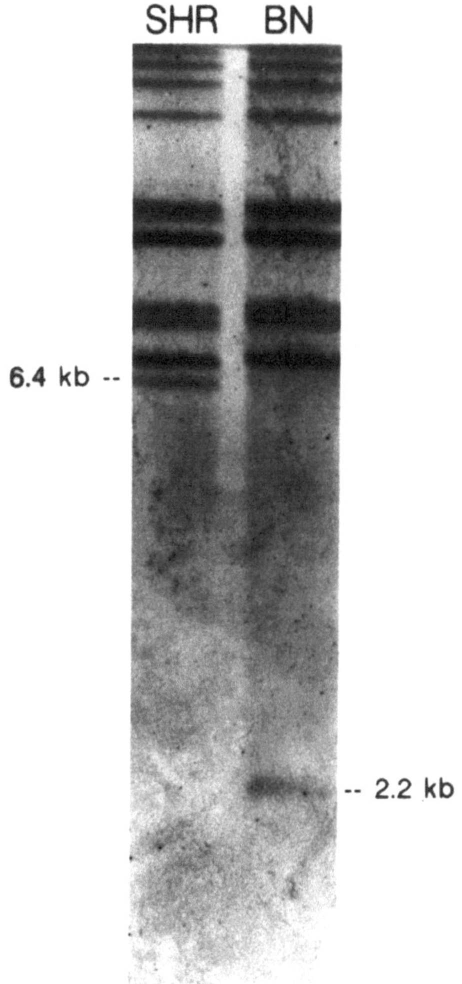

Fig. 3. Southern blot analysis depicting the RFLP used to distinguish the kallikrein alleles of the SHR and BN rat progenitor strains. The RFLP was detected by probing *Nde*I digested genomic DNA with a nonisotopically labeled cDNA probe encoding the 3′ end of the messenger RNA for rat pancreatic kallikrein

leukocyte antigen (HLA; major histocompatibility complex) haplotypes. Based on this observation, the authors concluded that a gene contributing to the pathogenesis of hypertension might be located within or near the major histocompatibility complex. In the recombinant inbred strains derived from the SHR and BN rat, Pravenec et al. (1989) recently studied the relationship between blood pressure and the pattern of inheritance of the RT1 major histocompatibility complex. The mean blood pressure of the RI strains that inherited the RT1 complex of the SHR (136 ± 3 mmHg) was found to be significantly greater than that of the RI strains that inherited the RT1 complex of the BN rat (124 ± 2 mmHg, $p < 0.001$; mean arterial pressures). These investigators also measured blood pressure in a congenic strain that differs from the SHR primarily at the RT1 complex and a short length of associated chromosome. The congenic strain was created by repeatedly backcrossing (N = 14) the RT1 complex of the BN rat onto the genetic background of the SHR. The blood pressure of the resulting SHR.1N congenic strain carrying the RT1 complex of the BN rat was found to be significantly lower than that of the SHR. This finding is consistent with the possibility that sequence variation in genes within or near the major histocompatibility complex (e.g., class I genes, class II genes, or other genes such as those coding for heat shock protein, 21 hydroxylase, or tumor necrosis factor) may have the capacity to affect blood pressure.

"Reverse" Genetics Approach to Gene Mapping in SHR

In recombinant inbred strains or in segregating F2 or backcross populations, one can search for chromosome regions that contain genes regulating blood pressure by studying the relationship between blood pressure and the pattern of inheritance of multiple polymorphic markers dispersed throughout the genome. Because this approach involves the testing of polymorphic markers from every chromosome, it does not require the investigator to make assumptions as to which genes are most likely to be involved in the pathogenesis of hypertension.

In the SHR, the major obstacle to using a "reverse" genetics approach to search for genes regulating blood pressure is the fact that only a modest number of polymorphic markers have been mapped in the rat (and even fewer have been tested in the SHR). Because the genome of the mouse is much better mapped than that of the rat, one might propose using a "reverse" genetics approach in spontaneously hypertensive mice (SHM) rather than the SHR. However, linkage studies of hypertension in the mouse are problematic because in this species the questionable, indirect tail-cuff technique is typically employed to measure blood pressure (Schlager 1974; Bunag 1983). For linkage studies of qualitative and quantitative traits, accurate phenotyping is absolutely critical. Although it is possible to directly measure blood pressure through arterial cannulae in unanesthetized, unrestrained mice, this direct approach is difficult to perform and has not been sufficiently validated in this species.

Another major obstacle to using SHM is the possibility that, in these models, as many as 20 genes may be involved in the pathogenesis of increased blood pressure (Schlager 1981, 1984). Because of the substantial non-genetic variability in blood pressure, it will be extraordinarily difficult to detect genes that have only minor effects on blood pressure. In the SHR, it has been estimated that just a few major genes may be involved in the pathogenesis of hypertension (Tanase et al. 1970). Although such estimates are to be viewed with caution (Kurtz et al. 1990b), it is noteworthy that, during the development of the SHR, severe hypertension appeared to have become fixed within 3–4 generations of brother × sister mating; further inbreeding resulted in only minor additional increases in blood pressure (Okamoto and Aoki 1963). For a complex quantitative trait (like blood pressure), this type of observation strongly suggests that the phenotype of interest is largely determined by just a few major genes (Plomin 1990).

Recent advances in molecular biology and rat genetics now make it possible to develop a genetic linkage map in the rat that can be applied to the study of spontaneous hypertension. As discussed by Lander and Botstein (1989), interval mapping using a complete map of gene markers can be used for the identification of chromosome regions that contain genes regulating blood pressure. To minimize the number of markers that must be initially tested when searching for QTLs, Chakravarti and Majumder (1989) proposed that a 50 cM map would represent a reasonable starting point. With such a map, any marker will show at most 23% recombination with a QTL (Chakravarti and Majumder 1989). Clearly, the likelihood of detecting linkage with a QTL regulating blood pressure will also depend on the magnitude of the blood pressure effect of the QTL. This serves to underscore the importance of studying an experimental model like the SHR in which just a few major genes may be determining of hypertension.

Assuming that the rat genome is about 1500 cM (Lander and Botstein 1989), one can grossly estimate the number of polymorphic markers required to begin developing a 50 cM linkage map (Lange and Boehnke 1982). If the available markers were evenly spaced, one would require a total of only 30 markers. While it is not feasible to directly develop a panel of evenly spaced polymorphic markers, it is possible to concentrate on identifying markers for genes with known chromosomal locations (from somatic cell hybrid or in situ hybridization studies) or with suspected chromosomal locations (from comparative mapping studies). Thus, even though more than 30 markers will be required to develop a 50 cM map, the total number should be less than that required if the markers were to be developed at random. Recently, Lathrop and colleagues have developed over 100 polymorphic markers in an SHR-WKY system by targeting hypervariable tandem repeat sequences (M. Lathrop, personal communication).

Identification of Polymorphic Markers for Gene Mapping in SHR

At the time of this writing, we have tested over 70 polymorphic markers in the recombinant inbred strains derived from the SHR and the normotensive BN rat.

Tables 1 and 2 depict the strain distribution patterns for some of the polymorphic markers tested to date. Based on an analysis of these strain distribution patterns, it appears that many of the markers we have tested are not tightly linked. Twenty-four of these markers have been assigned to nine rat linkage groups (of these linkage groups, six have been given chromosome assignments). Comparisons of strain distribution patterns of alleles shown in Tables 1 and 2 revealed several new linkages:

1. ES-6, LX, and RT5 loci are on linkage group (LG) VII;
2. the RMS15-2 and RCAC5-1 loci are closely linked to the RT2 locus and other markers of LGV;
3. the renin and FH-1 genes are linked on chromosome 13;
4. the RT8 and RMS15-6 loci are closely linked;
5. the RMS15-1 and RMS15-5 loci are closely linked;
6. the RMS15-4 and RCAC5-2 loci are closely linked; and
7. the RMS15-3 and RCAC5-7 loci are closely linked.

The chromosome/LG assignments of the last four cases remain to be established. Because RI strains are not efficient for detecting linkage with recombination fractions >0.12, we are mapping the unassigned polymorphic markers by linkage analysis of an F2 population and by somatic cell hybrid analysis.

We have been able to detect polymorphic markers with little difficulty simply by using standard Southern analysis, polymerase chain reaction (PCR) analysis, isoenzyme analysis by routine electrophoresis, and immunogenetic techniques. While it is feasible to use standard PCR and Southern blot techniques to detect polymorphic DNA markers for specific rat genes (particularly if one targets tandem repeat sequences), the use of newer techniques such as single strand conformation polymorphism (SSCP) analysis should greatly facilitate mapping efforts (Cawthon et al. 1990; Orita et al. 1989). This technique has the advantage of detecting DNA polymorphisms and point mutations at a variety of positions in DNA fragments, not just at enzyme recognition sites or tandem repeat sequences. Because DNA polymorphisms may occur every few hundred nucleotides in complex mammalian genomes, the method may provide a virtually unlimited supply of polymorphic markers.

In addition to developing polymorphic markers for specific rat genes, we have also targeted anonymous single locus and multi-locus VNTR sequences in our mapping effort. For example, by probing with multi-locus minisatellite and microsatellite probes, we have found extreme restriction fragment length polymorphism between SHR and BN rats. For the first two probes that we have tested, we have identified restriction fragments marking over 10 distinct polymorphic loci (as judged by the strain distribution patterns of the fragments; see RCAC5 and RMS15 loci in Tables 1 and 2 and Fig. 4 and 5). Although there is evidence that minisatellite sequences may be preferentially associated with telomeres in humans, a number of minisatellite and microsatellite fragments appear to be distributed over multiple chromosomes in rodents (at least in the mouse; Schafer et al. 1986; Zischler et al. 1989; Julier et al. 1990). We have also found that probes used to detect single locus VNTRs in humans (Nakamura et

Table 1. Strain distribution patterns of some polymorphic genetic markers

RI Strain	ALP-1	ANG	RCAC5-1	RCAC5-2	RCAC5-3	RCAC5-4	RCAC5-5	RCAC5-6	RCAC5-7	RCAC5-8	RCAC5-9	ES-2	ES-3	ES-6	ES-14	FH-1	IGK-1	Br
1A	B	H	H	B		H	H	H	B	B	B	H	H	H	H	H	B	
2A	B	H	H	B	H	H	H	B	B	B	H	H	H	H	H	H	B	
3A	B	H	H	B	H	B	H	H	B	B	H	H	H	H	H	H	H	H
4A	B	H	H	H	B	B	H	H	H	B	H	H	H	H	B	H	B	H
5A	H	B	B	H	B	B	H	H	B	H	B	B	B	B	B	B	H	H
7A	B	H	H	H	H	H		B	B	B	H	B	B	B	B	B	B	H
9A	H	B	B	B	H	H	H	H	H	H	B	B	B	H	B	B	B	H
10A	H	B	H	H	H	H	H	H	B	H	H	B	B	B	H	H	B	
13A	H	B	B	B	H	H	H	H	H	H	B	B	B	H	H	H		B
14A	H	B	H	B	B	H	H	H	H	H	H	H	H	H	H	H	H	
15A	B	H	B	H	H	H	B	B	H	B	H	B	B	B	B	H	B	H
16A	B	B	H	B	B	H	H	H	B	B		H	H	B	H	B	H	
17A	H	H	B	B	H	B	B	B	H	H	H	H	H	H	H	H	H	B
18A	B			H	H	H	H	H	B	H	H	B	B	H	B	H	H	H
19A	B	B	B	H	B				B			H	H	B	H	H	B	H
20A	B	H		B	H	H	H	B	H	B	B	H	H	B	B	H	B	H
21A	B	B	H	B	B	B	H	H	H	B	B	H	H	B	H	H	B	B
22A	B	B	H	B	H	H	B	B	B	H	H	H	H	H	H	B	B	H
23A	H	H	B	B	B	H	B	B	B	B	H	B	B	H	B	B	H	H
24A	H	B	B	B	B	B	B	B	B	H	B	B	B	H	B	H		H
25A	H	B	B	B	B	B	H	B	H	H	H	H	H	H	H	B	H	H
26A	B	B	H	H	B	H	B	B	H	H	B	H	H	H	H	B		
27A	H	B	H	B	B	H	H	H	B	B	B	H	H	B	H	B	H	B
29A	B	H	B	H	B	H	B	B	H	B	B	B	B	B	B	H	H	B
30A	B	B	B	H	H			H	H	H		B	B	H	B	H		H
31A	B	H	H	H	H	B	H	B	H	H	H	H	H	H	B	H	B	H
2C	B	B	H	B	B	H	B	B	B	H	H	H	B	B	H	H	H	B
02C	B	B	B	B	B	H	B	B	H	H	H	H	H	B	H	H	H	B
3C	B	B	B	H	H	H	H	B	B	B	H	B	B	B	B	H	B	B
5C	B	H	B	B	H	B	H	H	H	B	B	B	B	B	B	H	H	B
6C	B	B	H	H	B	B	H	B	H	H	H	H	H	B	H	B	B	H
8C		H	H	H	B	B	H	B	H		H	H	H	B	H	H	B	H
9C	H	B	H	B	B	B	H	B	H	B	B	B	B	B	B	H	B	B
10C	H	B	B	H	B	H	H	B	H	H	B	B	B	B	B	B	B	H
11C	B	B	B	B	B	B	H		H	B	H	B	B	B	H	H	H	
12C	H	B	B			B	H		H		H	B	B	B	B	B	H	H
13C	H	B	B			B	H		H			B	B	B	H	H	H	B

H, fixation of the SHR allele; B, fixation of the BN allele; ALP-1, alkaline phosphatase; ANG, angiotensinogen; RCAC5-1, microsatellite loci generated by probing *Hin*fI-digested DNA with a $(CAC)_5$ oligonucleotide; ES, esterase loci; FH-1, fumarate hydratase; IGK-1, immunoglobulin kappa chain; Br, brown coat color

Table 2. Strain distribution patterns of some polymorphic genetc markers. "H" designates fixation of the SHR allele, "B" fixation of the BN allele. GST-1, glutathione S-transferase (Matsumoto et al., 1983); IGH-1, immunoglobul in heavy chain; KAL, kallikrein gene family; LX, polydactyly-luxate syndrome; H, hooded; RMS 15-1-6, minisatellite loci generated by probing AluI digested DNA with oligonucleotide: AGAGGTGGGCAGGTG-GAGAGGTGGGCAGGTGG; REN, renin; RT1, major histocompatibility complex; RT2, RT5, RT8, cell surface alloantigens; SVP-1, seminal vesicle protein; C, albino coat color.

	GST-1	IGH-1	KAL	LX	H	RMS15-1	RMS15-2	RMS15-3	RMS15-4	RMS15-5	RMS15-6	REN	RT1	RT2	RT5	RT8	SVP-1	C
1A	H	H	B	H		H	H	B	B	H	H	H	B	H	H	H	B	H
2A	B	B	H	H		H	H	H	B	H	H	H	H	H	H	H	B	H
3A	H	H	H	H	H	H	H	H	B	H	B	H	H	H		B	H	B
4A	H	B	H	B		H	H	H	H	H	H	H	B		B	H	B	H
5A	H	H	H	B		H	B	B	H	H	H	B	H	B		H	H	H
7A	B	B	H	B	B	H	H	B	H	H	B	B	B	B		B	B	B
9A	B	B	B	H	H	B	B	B	B	B	H	B	B	B		H	B	B
10A	H	B	B	B		H	H	B	H	H	H	H	B	B		H	B	H
13A	H	H	H	H	B	H	B	H	B	H	H	H	B	B		H	H	B
14A	H	H	B	H		H	H	B	B	H	H	H	B	H		H	H	H
15A	H	B	B	H	H	H	B	H	H	B	B	H	H	B	H	B	B	B
16A		H	H	B		H	H	B	B	H	B		H	H		B	H	H
17A		B	H	H	B	H	H	B	B	H	H	H	H	H	H	H	H	B
18A	H	B	H	H	B	H	H			B	B	H	B	H	H	B	H	B
19A	B	B	B	H	H	H	H	B	H	H	B	B	B	B		B	B	B
20A	H	H	H	B	H							H	H		B	B	B	B
21A	H	B	B	B	B	B	H	H	B	B		H	H	H	B	H	B	H
22A	B	H	B	H	H	H	H	B	B	H	H	H	B	H		H	H	B
23A	H	H	B	H	H	B	B	B	B	B	B	H	B	B	H	H	B	B
24A	H	B	B	H	H	H	B	B	B	H	B	B	B	B	H	B	B	B
25A	H	H	B	B	H	H	B		B	H	H	B	B	B	B	H	B	B
26A	B		B	H		B	H	B	H	B	B	B	B	H		B	B	H
27A	H		B	B	B	B	H	B	B	B	H	B	B	H		H	B	B
29A	B	B	H	B	B	B	B	B	H	B	B	H	B	B		H	B	B
30A	H		B	H	H	H	B	H	H	H	B	B	H	B	H	H	B	B
31A	B	B	B	H	H	H	H	H	H	H	H	H	H	H		H	B	B
2C		H	B	B	B	B	B	H	B	B	H	H	H	B	B	H	B	B
02C	B		B	B	B	B	H	B	B	H	H	H	B	H	B	B	B	B
3C	B		B	B	H	B	B	B	H	B	H	B	B	B	B	B	H	B
5C	H	H	H	B	B	B	B	H	B	B	H	H	H	B	B	H	H	B
6C	B	B	H	B	B	B	H	B	H	B	H	B	H	H		H	H	B
8C	H	H	B	B	H	B	H	H	H	B	B	H	B	H		B	H	B
9C	H	B	H	B	B	B	H	H	B	B	H	B	B	H	B	H	B	B
10C	H		B	B	H	B	B	H	H	B	B	B	B	B	B	B	H	B
11C	H	H	H	B		B	B	H	B	B	H	H	H	B	B	H	B	H
12C	H	H	H	B	B	B	B	H		B	H	H	H	B	B	H	H	B
13C	B	H	H	B	H	B	B	H		B	H	H	B	B	B	H		B

al. 1987) can be used to detect RFLPs in the rat (Fig. 6). Recently, Julier et al. (1990) constructed primary minisatellite genetic linkage maps in the mouse by testing human VNTR probes in a large panel of RI strains derived from C57BL/6J and DBA/2J mice.

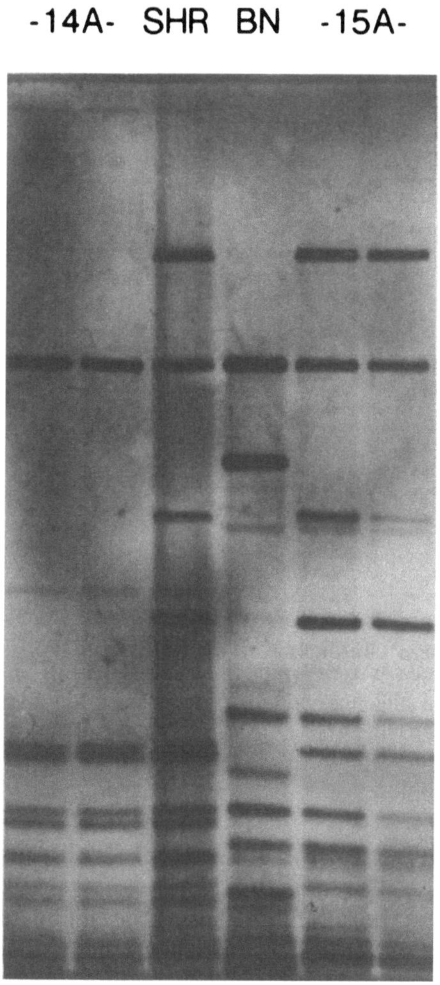

Fig. 4. Southern blot analysis of *Hin*fI digested genomic DNA from an SHR, a BN rat, and two recombinant inbred (RI) strains (14A and 15A, two rats per RI strain) generated by probing with a nonisotopically labeled microsatellite repeat sequence $(CAC)_5$. In Table 1, the loci identified by this probe-enzyme combination have been provisionally designated by the prefix "RCAC5"

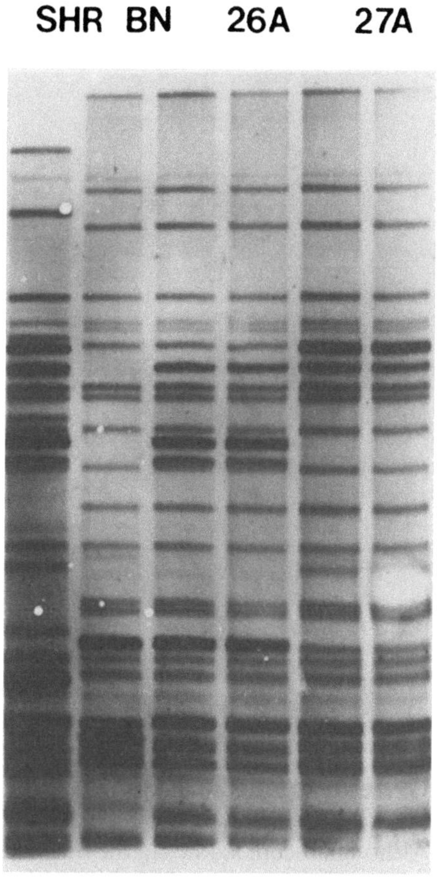

Fig. 5. Southern blot analysis of *Alu*I digested genomic DNA from an SHR, a BN rat, and two recombinant inbred –RI) strains (26A and 27A, two rats per RI strain) generated by probing with a nonisotopically labeled oligonucleotide corresponding to the consensus repeat sequence of the human myoglobin 33.15 minisatellite (Jeffreys et al. 1985). In Table 2, the loci identified by this probe-enzyme combination have been provisionally designated by the prefix "RMS15"

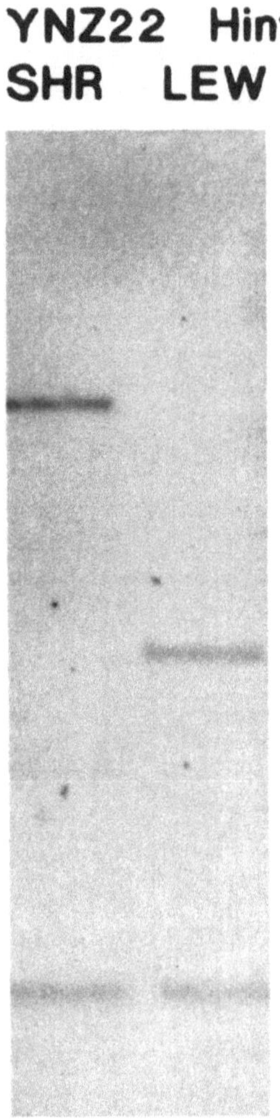

Fig. 6. Southern blot analysis of an RFLP in SHR and Lewis rats generated by probing *Hin*fI-digested genomic DNA with an oligonucleotide corresponding to the core sequence of the human YNZ22 variable number of tandem repeats sequence (Nakamura et al. 1987)

Gene Mapping in Experimental Hypertension – General Problems

Although recent studies in the SHR suggest the feasibility of using the candidate gene and reverse genetics approaches to search for chromosome regions that contain genes regulating blood pressure, some general problems require further consideration.

Blood Pressure Phenotyping

If one is to perform a genetic study of a qualitative or quantitative trait, it is essential that each subject's phenotype be accurately characterized. Indeed, it has been suggested that some of the problems involved in reproducing linkage studies of mental illness may in part be a consequence of difficulties in accurately determining phenotype (Barnes 1989). Problems in obtaining accurate BP measurements may pose the greatest obstacle to successful linkage studies in human and experimental hypertension.

Although many investigators studying hypertension in rats (and mice) employ indirect tail-cuff plethysmography, the validity of this technique is suspect because the method not only requires restraint of the animal or anesthesia, but also the induction of an increase in body temperature (Bunag 1983; Ferrari et al. 1990). Furthermore, the tail-cuff method typically provides for only a few measurements of systolic blood pressure; accurate measurement of diastolic pressure is not possible. Blood pressure is a highly labile trait that can fluctuate greatly over short periods of time. Therefore, it would seem essential to average multiple measurements of blood pressure when attempting to assign a blood pressure value to a given individual.

In the study in SHR and Lewis rats in which we tested for linkage between blood pressure and an RFLP marking the renin gene, we estimated blood pressure by obtaining the computerized signal averages of hundreds of direct measurements of pressure in the unanesthetized, unrestrained state (Kurtz et al. 1990a). However, these measurements of blood pressure were obtained over a period of just several hours and may not accurately represent blood pressure over the entire day. It should also be noted that, in the studies with the recombinant inbred strains, the direct measurements of blood pressure were performed in anesthetized animals (Pravenec et al. 1989). In future studies in the RI strains, it will be important to obtain prolonged measurements of blood pressure in conscious, unrestrained animals. The optimal approach would be the use of radiotelemetry devices that allow for continuous, direct measurements of arterial pressure. At the present time, however, this approach is very expensive and time-consuming.

Statistical Considerations

In the standard linkage analysis of a genetic marker and a qualitative trait, a lod score of 3 (corresponding to a critical value of approximately 0.001) is generally required for an investigator to strongly conclude that linkage is present. Because the prior probability that two randomly chosen genes will be linked is rather low, the critical threshold is set at 0.001 to yield an actual false positive rate of approximately 0.05 (assuming a reasonable power of detecting linkage when it is present; Chotai 1984). In studies that involve a single "candidate" gene marker and a quantitative trait with multiple genetic components, choice of an appropriate critical threshold is problematic. With some candidate gene markers, the prior evidence for linkage to a blood pressure regulatory locus may be reasonably strong. In this circumstance, it may be appropriate to set a less stringent critical value for the linkage analysis. The existence of multiple blood pressure regulatory loci would also increase the prior probability of detecting linkage (with either a candidate gene marker or an anonymous gene marker). On the other hand, when many polymorphic markers are tested, the probability of a false positive result increases and stringent critical values are clearly required (Lander and Botstein 1989). Because of the uncertainty regarding appropriate critical thresholds for genetic studies of blood pressure, it would seem prudent to independently confirm any "positive" results before drawing strong conclusions about possible chromosome locations of QTLs regulating blood pressure.

Selection of Progenitor Strains

In setting up classical crosses to generate F2, backcross, and RI populations, it is important that the progenitor strains employed be highly inbred and therefore homozygous at virtually all loci. If not, the mapping studies wil be highly problematic. In most experimental studies in rodents, it is often assumed that the strains employed are fully inbred. However, most commercial and laboratory breeders maintain little or no genetic quality control of their stocks and marked genetic variability within "inbred" strains is not uncommon (Cramer 1983; Festing 1982). In a recent DNA fingerprint study in the WKY rat, the most commonly employed control for the SHR, we found molecular evidence of genetic variability among WKY from different commercial facilities as well as among WKY within a single breeding facility (Kurtz et al. 1989). As a consequence of these findings, we have avoided use of WKY rats in genetic studies.

Consideration of Sex-linked Effects on Blood Pressure

In a recent genetic study in rats in which blood pressure was measured by the indirect tail-cuff technique, it was suggested that the Y chromosome of the SHR might be an important determinant of the full expression of hypertension (Ely and Turner 1990). However, a preliminary analysis of the blood pressure data from the HXB and BXH sets of RI strains has not revealed any effects of the SHR Y chromosome on blood pressure. Although the proposed Y chromosome effect remains to be confirmed, investigators should recognize the possibility that detection of linkage with QTLs regulating blood pressure might be facilitated by the use of male SHR in the breeding protocols. There is little evidence suggesting that the X chromosome carries major genes determining of hypertension in SHR (Tanase 1979; Ely and Turner 1990).

Prospects for Fine Mapping

Although a "reverse" genetics approach using a dense (10–20 cM) linkage map may enable one to estimate the relative position of a QTL within an interval, it does not allow one to pinpoint the precise location of a QTL regulating blood pressure. In plants, QTLs have been mapped to small intervals using the technique of substitution mapping; however, such approaches may not be feasible for QTL mapping in animals (Paterson et al. 1990). Nevertheless, the identification of even a large interval that contains a QTL regulating BP would represent a major advance in the genetics of hypertension. The identification of specific chromosome regions that contain genes regulating blood pressure would provide direction in selecting appropriate candidate genes for further study. At present, candidate genes for hypertension are targeted primarily on theoretical grounds alone (i.e., without evidence for genetic linkage to the regulation of blood pressure). However, blood pressure is a highly complex trait and it is difficult to select among the hundreds of potential candidate genes. Objective linkage data are required to enable investigators to rationally narrow the focus on appropriate candidate genes for hypertension. Finally, the identification of chromosome regions that contain genes regulating blood pressure will allow for the development of congenic strains that may be used to study the biochemical and physiologic determinants of hypertension. Ultimately, however, techniques such as gene targeting and homologous recombination will be required to determine the precise nature of the genetic events that underlie hypertension.

References

Barnes FM (1989) Troubles encountered in gene linkage land. Science 243: 313–314

Bunag RD (1983) Facts and fallacies about measuring blood pressure in rats. Clin Exp Hypertens [A] 5 (10): 1659–1681

Cawthon RM, Weiss R, Xu G, Viskochil D, Dunn D, Gesteland R, O'Connell P, White R (1990) A major segment of neurofibromatosis type 1 gene: cDNA sequence, genomic structure, and point mutations. Cell 62: 193–201

Chakravarti A, Majumder PP (1989) Strategies for mapping traits using a genome map. Cold Spring Harbor Symp Genome Mapping 32

Chotai J (1984) On the lod score method in linkage analysis. Ann Hum Genet 48: 359–378

Cramer DV (1983) Genetic monitoring techniques in rats. ILAR News 16: 15–20

Ely DL, Turner ME (1990) Hypertension in the spontaneously hypertensive rat is linked to the Y chromosome. Hypertension 16: 277–281

Ferrari AU, Daffonchio A, Albergati F, Berotoli P, Mancia G (1990) Intra-arterial pressure alterations during tail-cuff blood pressure measurements in normotensive and hypertensive rats. J Hypertens 8: 909–912

Festing MFW (1982) Genetic contamination of laboratory animal colonies: an increasingly serious problem. ILAR News 15: 6–8

Frohlich ED, Trippodo NC (1981) Similarities of genetic (spontaneous) hypertension: man and rat. Circ Res 48: 309–319

Gerbase-DeLima M, Delima JJG, Persoli LB, Bernardes Silva H, Marcondes M, Bellotti G (1989) Essential hypertension and histocompatibility antigens. Hypertension 14: 604–609

Gray SD (1984) Pressure profiles in neonatal spontaneously hypertensive rats. Biol Neonate 45: 25–32

Guidi E, Hollenberg NK (1987) Differential pressor and renal vascular reactivity to angiotensin II in spontaneously hypertensive and Wistar-Kyoto rats. Hypertension 9: 591–597

Jeffreys AJ, Wilson V, Thein SL (1985) Hypervariable minisatellite regions in human DNA. Nature 314: 67–73

Julier C, De Gouyon B, Georges M, Gunet J-L, Nakamura Y, Avner P, Lathrop GM (1990) Minisatellite linkage maps in the mouse by cross-hybridization with human probes containing tandem repeats. Proc Natl Acad Sci USA 87: 4585–4589

Kurtz TW, Morris RC (1987) Biological variability in Wistar-Kyoto rats: implications for research with the spontaneously hypertensive rat. Hypertension 10: 127–131

Kurtz TW, Montano M, Chan L, Kabra P (1989) Molecular evidence of genetic heterogeneity in Wistar-Kyoto rats: implications for research with the spontaneously hypertensive rat. Hypertension 13: 188–192

Kurtz TW, Simonet L, Kabra PM, Wolfe S, Chan L, Hjelle BL (1990a) Cosegregation of the renin allele of the spontaneously hypertensive rat with an increase in blood pressure. J Clin Invest 85: 1328–1332

Kurtz TW, Casto R, Simonet L, Printz M (1990b) Biometrical genetic analysis of blood pressure in the spontaneously hypertensive rat. Hypertension 16: 718–724

Lander Es, Botstein D (1989) Mapping Mendelian factors underlying quantitative traits using RFLP linkage maps. Genetics 121: 185–199

Lange K, Boehnke M (1982) How many polymorphic genes will it take to span the human genome? Am J Hum Genet 34: 842–845

Lindpainter K, Takahashi S, Ganten D (1990) Structural alterations of the renin gene in stroke-prone spontaneously hypertensive rats: examination of genotype-phenotype correlations. J Hypertens 8: 763–774

Matsumoto K, Gasser DL (1983) Polymorphism of glutathione S-transferase in laboratory rats. Biochem Genet 21: 1209–1251

Mori M, Ishizaki I, Yamada T, Chen H, Sugiyama T, Serikawa T, Yamada J (1989) Restriction fragment length polymorphism of the angiotensinogen gene in inbred rat strains and mapping of the gene on chromosome 19q. Cytogenet Cell Genet 50: 42–45

Mullins JJ, Peters J, Ganten D (1990) Fulminant hypertension in transgenic rats harbouring the mouse Ren-2 gene. Nature 344: 541–544

Nakamura Y, Leppert M, O'Connell P, Wolff R, Holm T, Culver M, Martin C, Fujimoto E, Hoff M, Kumlin E, White R (1987) Variable number of tandem repeat (VNTR) markers for human gene mapping. Science 235: 1616–1622

Ohkubo J, Kawakami H, Kakehi Y, Takumi T, Arai H, Yokota Y, Iwai M, Tanabe Y, Masu M, Hata J, Iwao H, Okamoto H, Yokoyama M, Nomura T, Katsuki M, Nakanishi S (1990) Generation of transgenic mice with elevated blood pressure by introduction of the rat renin gene and angiotensinogen genes. Proc Natl Acad Sci USA 87: 5133–5157

Okamoto K, Aoki K (1963) Development of a strain of spontaneously hypertensive rats. Jpn Circ J 27: 282–293

Orita M, Iwahana H, Kanazawa H, Hayashi K, Sekiya T (1989) Detection of polymorphisms of human DNA by gel electrophoresis as single-strand conformation polymorphisms. Proc Natl Acad Sci USA 86: 2766–2770

Paterson AH, DeVerna JW, Lanini B, Tanksley SD (1990) Fine mapping of quantitative trait loci using selected overlapping recombinant chromosomes, in an interspecies cross of tomato. Genetics 124: 735–742

Plomin R (1990) The role of inheritance in behavior. Science 248: 183–188

Pravenec M, Klir P, Kren V, Zicha J, Kunes J (1989) An analysis of spontaneous hypertension in spontaneously hypertensive rats by means of new recombinant inbred strains. J Hypertens 7: 217–222

Pravenec M, Simonet L, Kren V, Kunes J, Levan G, Szpirer J, Szpirer C, Kurtz T (1991a) The rat renin gene: assignment to chromosome 13 and linkage to the regulation of blood pressure. Genomics 9: 466–472

Pravenec M, Kren V, Kunes J, Scicli AG, Carretero OA, Simonet L, Kurtz TW (1991b) Cosegregation of blood pressure with a kallikrein gene family polymorphism. Hypertension 17: 242–246

Rapp JP (1983a) Genetics of experimental and human hypertension. In: Genest J, Kuchel O, Hamet P, Cantin M (eds) Hypertension, pathophysiology and treatment. McGraw-Hill, New York, pp 582–598

Rapp JP (1983b) A paradigm for identification of primary genetic causes of hypertension in rats. Hypertension 5 [Suppl I]: I-198–I-203

Rapp JP (1987) Use an misuse of control strains for genetically hypertensive rats. Hypertension 10: 7–10

Rapp JP, Wang S-M, Dene H (1989) A genetic polymorphism in the renin gene of Dahl rats cosegregates with blood pressure. Science 243: 542–544
Samani NJ, Brammar WJ, Swales JD (1989) A major structural abnormality in the renin gene of the spontaneously hypertensive rat. J Hypertens 7: 249–254
Schafer R, Sher A, Epplen JT (1986) The organization of the evolutionarily conserved GATA/GACA repeats in the mouse genome. Chromosoma 93: 502–510
Schlager G (1974) Selection for blood pressure levels in mice. Genetics 76: 537–549
Schlager G (1981) The genetically hypertensive mouse. In: Worcel M (ed) Trends in arterial hypertension. North-Holland, New York, pp 321–331
Schlager G (1984) Genetic hypertension in different strains of mice. In: De Jong W (ed) Handbook of hypertension: experimental and genetic models of hypertension. North-Holland, New York, pp 192–199
Tanase H (1979) Genetic control of blood pressure in spontaneously hypertensive rats. Exp Anim 28: 519–530
Tanase H, Suzuki Y, Ooshima A, Yamori Y, Okamoto K (1970) Genetic analysis of blood pressure in spontaneously hypertensive rats. Jpn Circ J 34: 1197–1212
Weber MA (1988) Whole-day blood pressure. Hypertension 11: 288–298
Williams GH, Braley LM, Menachery A (1982) Decreased adrenal responsiveness to angiotensin II: A defect present in spontaneously hypertensive rats. J Clin Invest 69: 31–37
Woodley-Miller C, Chao J, Chao L (1989) Restriction fragment length polymorphisms mapped in spontaneously hypertensive rats using kallikrein probes. J Hypertens 7: 865–871
Zischler H, Nanda I, Schafer M, Schmid M, Epplen JT (1989) Digoxigenated oligonucleotide probes specific for simple repeats in DNA fingerprinting and hybridization in situ. Hum Genet 82: 227–233

Transgenic Mice and the Development of Animal Models and Resources for Hypertension Research*

C. D. Sigmund, C. A. Jones, J. R. Fabian, C. Wu, C. M. Kane, M. K. Ellsworth, F. D. Pacholec, and *K. W. Gross*△

Summary

In identifying the genic lesions which characterize hypertension and other cardiovascular disorders, the mouse, with its wealth of classical and molecular genetic resources, has much to offer as an experimental system. Transgenic technology, in particular, affords exciting opportunities for devising new approaches to study complex physiological and developmental processes. Here we will describe studies in which transgenic technology is used to target neoplasia to renin-expressing cells as a prelude to isolating cell lines for in vitro study. Such cell lines afford the potential to study intracellular mechanisms regulating renin gene expression and secretion independent of complicating influences found in situ.

Introduction

The enzyme renin has long been known to play an important role in the regulation of extracellular volume and blood pressure. The classical site of renin synthesis and release in adult mammals is the juxtaglomerular (JG) cells of the afferent arteriole in the kidney, where expression and secretion are physiologically regulated by a variety of mechanisms. More recently compelling evidence for the expresion of renin in a spectrum of other tissues during ontogeny has accumulated (Naruse et al. 1983; Field et al. 1984; Field and Gross 1985; Sealy et al. 1986; Fabian et al. 1989; Ekker et al. 1989).

A major limitation in defining in detail the mechanisms of expression and regulation operating at the juxtaglomerular site and other sites has been the absence of suitable *in vitro* cell culture systems, retaining adequate differentiated character. A further complicating issue lies in the finding that genetic heterogeneity exists at the murine renin locus/loci (Piccini et al. 1982). The three

* This work was funded by NIH grants, numbers HL35792 and GM30248. This project was supported in part by BRSG SO7 RR-05648-23 awarded by the Biomedical Research Support Grant Program, Division of Research Resources, National Institutes of Health. C.D.S. is the recipient of NIH fellowship HL07963. △Corresponding author.

K. Berg, V. Bulyzhenkov, Y. Christen, P. Corvol (Eds.)
Genetic Approaches to Coronary Heart Disease
and Hypertension

mouse renin genes, *Ren-1*^{c}, *Ren-1*^{d} and *Ren-2,* exhibit a complicated pattern of gene-specific expression. Although all three murine renin genes retain a high degree of DNA sequence similarity, they have overlapping yet distinct tissue-specific expression profiles (Field and Gross 1985; Fabian et al. 1989; Ekker et al. 1989). The lack of renin-expressing cell lines has severely impeded definition of the *cis-acting* DNA sequence conferring renin expression at various organ sites.

To circumvent this limitation, a transgenic mouse approach has been implemented by a number of groups. Although time consuming, such an approach provides an assay for assessing effects in all tissues throughout the organism's development. Such studies employing genomic sequences for the mouse *Ren-2* gene have shown that important regulatory elements appear to be closely linked to the renin coding regions (Tronik et al. 1987; Mullins et al. 1988, 1989; Miller et al. 1989).

In our efforts to define such regulatory sequences more precisely and to characterize the renin-angiotensin system further, we have constructed transgenic mice with 5′ flanking sequences of the mouse *Ren-2* gene fused to a viral oncogene SV40 T antigen. These experiments have permitted us to exploit:

1. the sensitive reporter function of the Tag polypeptide product to assess the fidelity of the linked regulatory sequences in directing tissue-specific transcription in an animal system, and
2. the potent oncogenicity of T antigen in the target cells to isolate renin expressing cell lines for more detailed future in vitro studies.

Finally, profound pathophysiological effects have been found to attend expression of this fusion transgene, and they have been preliminarily assessed for additional insights into mechanisms for the maintenance of blood pressure and electrolyte homeostasis.

Methods

Production of Transgenic Mice

Construction of pR2 4.6 Tag has been previously described in detail (Sigmund et al. 1990a, b). The production of transgenic mice with this construct and their initial characterization has been published previously (Sigmund et al. 1990 a, b, c; 1991; Jones et al. 1990). The isolation of renin-expressing JG cell lines has been described in detail (Sigmund et al. 1990c).

Histology, Immunocytochemistry, In Situ Hybridization, and Electron Microscopy

Tissue samples were fixed in 10 % buffered formalin for paraffin embedding, and 5 μm sections were cut. For routine histological analysis, the tissue sections were stained with hematoxylin and eosin.

Indirect immunofluorescent assays to detect renin utilized a rabbit polyclonal antiserum produced to purified mouse submaxillary gland renin (a gift from R. Pratt, Stanford University). Cells were grown on coverslips and lightly fixed with 2% formalin in phosphate buffered saline (PBS) for 15 min. Cells were permeabilized with 0.1% Triton X-100 PBS for 15 min and washed in PBS. Anti-renin antiserum diluted 1:100 in 0.05% Triton X-100 PBS was then applied for 45 min followed by two 5 min washes. FITC conjugated donkey anti-rabbit IgG (Jackson ImmunoLabs) diluted 1:50 was then applied for 30 min followed by two 5 min washes. Coverslips were then mounted and examined on a BioRad MRC 600 Confocal Imaging System using an argon laser. Photomicrographs were made of a maximum projection of 0.5 μm thick-Z-series optical sections.

Indirect immunoperoxidase assays were performed as previously described on fetuses (Jones et al. 1990) and cells (Sigmund et al. 1990c). Fetal sections (Jones et al. 1990) or tissue culture cells (Sigmund et al. 1990c) were hybridized in situ to detect renin mRNA with an ^{35}S-labeled *Ren-2*^{d} cDNA. RNA probes were produced by in vitro transcription using SP6 RNA polymerase (Riboprobe, Promega Biotech, Madison, WI). The probe, pSLM, is a 911 base-pair antisense RNA complementary to bases 22 to 933 of the *Ren-2*^{d} mRNA. ^{35}S-labeled probes were reduced in size to 50–100 bases by alkaline hydrolysis at 60 °C (Lynn et al. 1983). The sense strand was transcribed and used as a negative control probe.

Cells in which renin was expressed at detectable levels displayed silver grains as detected by liquid emulsion autoradiography with Kodak NTB-2 emulsion diluted 1:1 with water. Slides were autoradiographed for 7–14 days at 4 °C and, after developing, were stained with hematoxylin and eosin Y to visualize tisue structure.

Ultrastructural analysis was performed by electron microscopy of the 4.1 cells as previously described (Sigmund et al. 1990c).

Analysis of Renin and Transgene Expression in the Cell Line 4.1

Primer Extension. The primer extension protocol described by Field and Gross (1985) was used to confirm expression of *Ren-1*^{c} in the 4.1 cell line, a cell line derived from a *Ren-1*^{c} mouse. Five μg of cell line RNA o 80 μg kidney RNA was hybridized with either a 38 mer oligonucleotide to distinguish *Ren-1* from *Ren-2* transcripts or a 37 mer to distinguish interspecific renin genes (mouse vs. human). Primers were extended in the presence of dATP, dGTP, TTP and ddCTP for the 38 mer and dCTP, dGTP, TTP and ddATP for the 37 mer.

Serum Response. The 4.1 cells were grown until 90%–100% confluency in DMEM supplemented with 15% fetal bovine serum (FBS). Media containing 15% FBS were added to one set of flasks and incubated for 48 h, DMEM without serum was added to a second set of flasks for 48 h. DMEM without serum was added to a third set of flasks for 48 h followed by the addition of DMEM supplemented with 15% FBS for 24 h. Ltk$^-$ cells were grown in DMEM supplemented with 10% calf serum. The cells were harvested by scraping the dish, and RNA was extracted from the cells as previously described (Sigmund et al. 1990c; Ausubel et al. 1987). For analysis of expression, the indicated amount of RNA (Fig. 6) was applied to nitrocellulose using a slot blot manifold. Strips were hybridized with antisense RNA probe to renin or with randomly primed DNA probed (Tag or actin). Films were scanned with the linear range of the film using a Helena Quick Scan.

Results

Transgenic Expression

Transgenic mice were produced with the fusion construct depicted in Fig. 1. It consists of 4.6 kb of the 5′ flanking region of *Ren-2*^{d} fused to the SV40 T antigen structural gene. Multiple founder lines expressing the transgene were established. In general, they appear to exhibit similar properties. Two lines designated T4 and T7 have been studied in the greatest detail. All lines were established and maintained on a *Ren-1*^{c} genetic background.

Expression of the transgene and the endogenous renin gene was examined and compared from mid-gestation through adulthood. The results obtained are described in detail in Jones et al. 1990 and Sigmund et al. (1990a, 1991). In brief, the transgene and endogenous renin gene exhibited the spatial and temporal patterns of expression appropriate to *Ren-2* and *Ren-1*^{c}, respectively. They thus exhibited coexpression at sites previously identified as such for *Ren-2* and *Ren-1*^{c}, such as kidney and fetal adrenal, and divergence at sites known to differ, e.g., anterior prostate and adult adrenal cortex (Field and Gross 1985; Fabian et al. 1989). In all cases, Northern blot and S_1 analysis indicated that appropriate transcription initiation sites were utilized.

An example of colocalization is shown in Fig. 2A–C for the developing vasculature of the fetal kidney and the cortex of the fetal adrenal gland. Expression of the endogenous renin gene *Ren-1*^{c} was detected by in situ

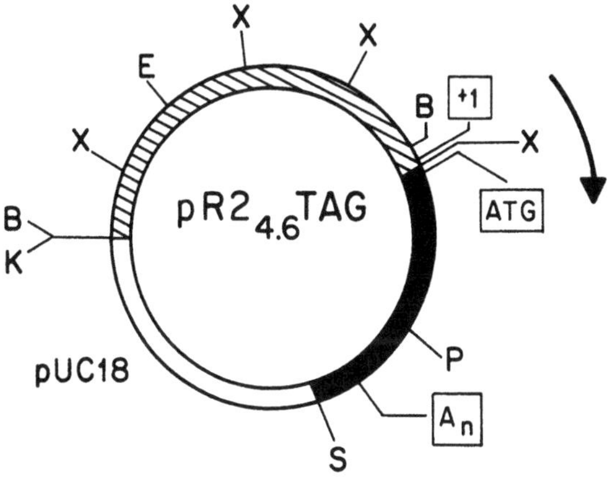

Fig. 1. Physical map of the Ren-Tag fusion construct. This construct contains 4.6 kb of 5′ flanking sequence from the mouse *Ren-2*^{d} gene (hatched) fused to the coding region of SV40 T antigen (filled). Transcription initiates at the major start site with the *Ren-2* sequence (+1) and translation begins with T antigen (ATG). The fusion construct was excised from the plasmid by digestion with *Kpn* 1 (K) and *Sal*I (S). The following sites are also identified: B, *Bam*HI; E, *Eco*RI; X, *Xba*I; P, *Pvu*II; A_n, poly A addition site

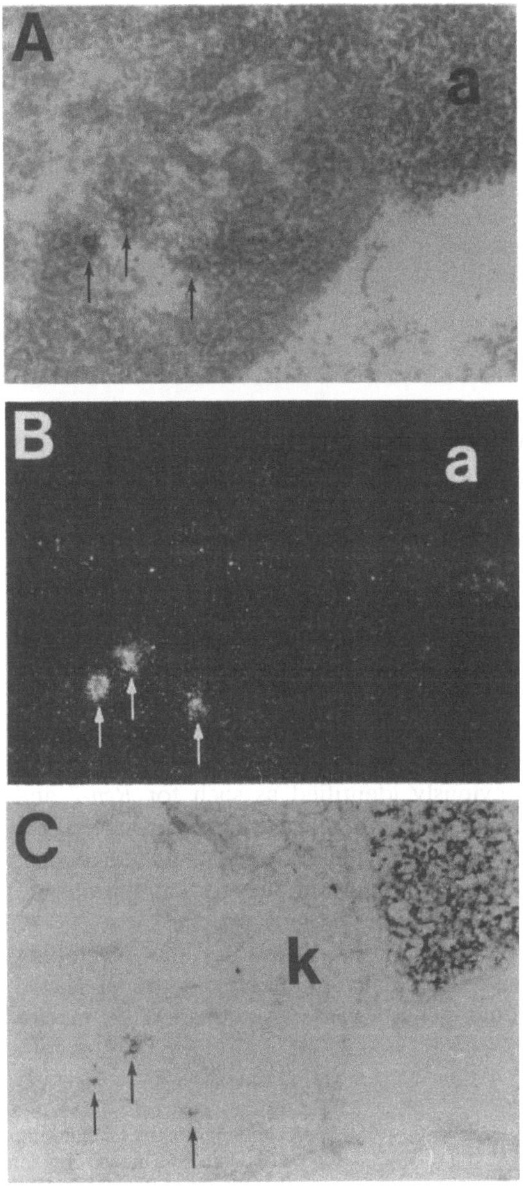

Fig. 2A–C. Colocalization of renin mRNA and SV40 large Tag on 10 μm serial sections of a 15-day post-coitum transgenic fetus. **A, B** Bright and dark field iluminations, respectively, of a fetal kidney and adrenal gland tissue autoradiography hybridized in situ with a probe to detect renin mRNA. **C** A bright field illumination of the serial section stained immunohistochemically to detect transgene product. *Arrows* indicate examples of renal vasculature; *k*, kidney; a, adrenal gland

hybridization in the differentiating smooth muscle cells (SMC) along the developing kidney vasculature and diffusely throughout the adrenal cortex. The Tag polypeptide, detected by indirect immunocytochemistry, serves as a reporter function for the *Ren-2* transcriptional regulatory sequences of the transgene construct. The immunoperoxidase staining colocalizes on adjacent sections with the in situ hybridization signal.

The transgenic animals exhibit the expected ontogenetic shift in the renin-expressing cell population (Sigmund et al. 1990a; Jones et al. 1990). Thus expression of the endogenous renin gene and the transgene is progressively restricted during late gestation and the neonatal period from SMC of the major kidney arteries to the classical site of adult expression, the epithelioid juxtaglomerular (JG) cell (Sigmund et al. 1990a; Jones et al. 1990). Moreover, during this time frame, it could be demonstrated that expression of both the transgene and *Ren-1*^{c} was responsive to modulation by angiotensin converting enzyme (ACE) inhibitors (Sigmund et al. 1991). Thus upstream renal vascular SMC can be recruited to express both renin and Tag.

Renal Vascular Pathology in the Ren-Tag Mouse

In addition to providing an exquisite reporter function, Tag is a potent oncogene. The consequences of this potent oncogenicity become evident throughout ensuing development, most notably in dramatic restructuring within the renal vascular bed through hyperplasia of SMC and the stochastic occurrence of overt tumors in the kidney and other known sites of renin expression.

As shown in Fig. 3A, B and described more fully in Sigmund et al. (1991), by 12-16 weeks marked proliferation of SMC is evident within the medial layer of the intra-renal arterial walls. This hyperplastic response is found in varying degrees of severity throughout the entire renal arterial tree, including the interlobar, arcuate, interlobular and afferent branches. The hyperplasia can be accompanied by mild to moderate secondary intimal fibrosis causing thickening of the arterial wall and, in some cases, virtual occlusion of the vessel lumen.

In such adult transgenic animals, T antigen-containing cells were are clearly evident scattered throughout the renal cortex. The expressing cells are localized to medial (not intimal) vascular SMC as well as occasional glomerular (mesangial) cells and solitary cells in the interstitium, possibly capillaries or associated cells. Areas of intra-renal cellular transformation, short of overt tumorigenesis, are also occasionally encountered.

Preliminary physiological analyses of the renal and cardiovascular status of these animals are described in Sigmund et al. (1991) and Jacob et al. (1991).

These studies indicate that the striking renal vascular pathology is associated with markedly suppressed levels of renal renin mRNA, renal renin content and plasma renin. At this stage, Northern blot analysis indicates that ACE inhibition can no longer elicit and increased renin response. The transgenic animals also exhibit plasma volume expansion. Surprisingly, despite the pronounced renal vascular pathology and volume expansion, these animals have been found to be normotensive (Sigmund et al. 1991; Jacob et al. 1991).

Isolation of a Renin-expressing Cell Line

Overt tumors formed stochastically in adult mice at known sites of renin expression. These include the submandibular gland, kidney, adrenal gland, subcutaneous tissue and testes (Sigmund et al. 1990b, c; Sigmund and Gross 1990). This finding was in complete agreement with our data demonstrating

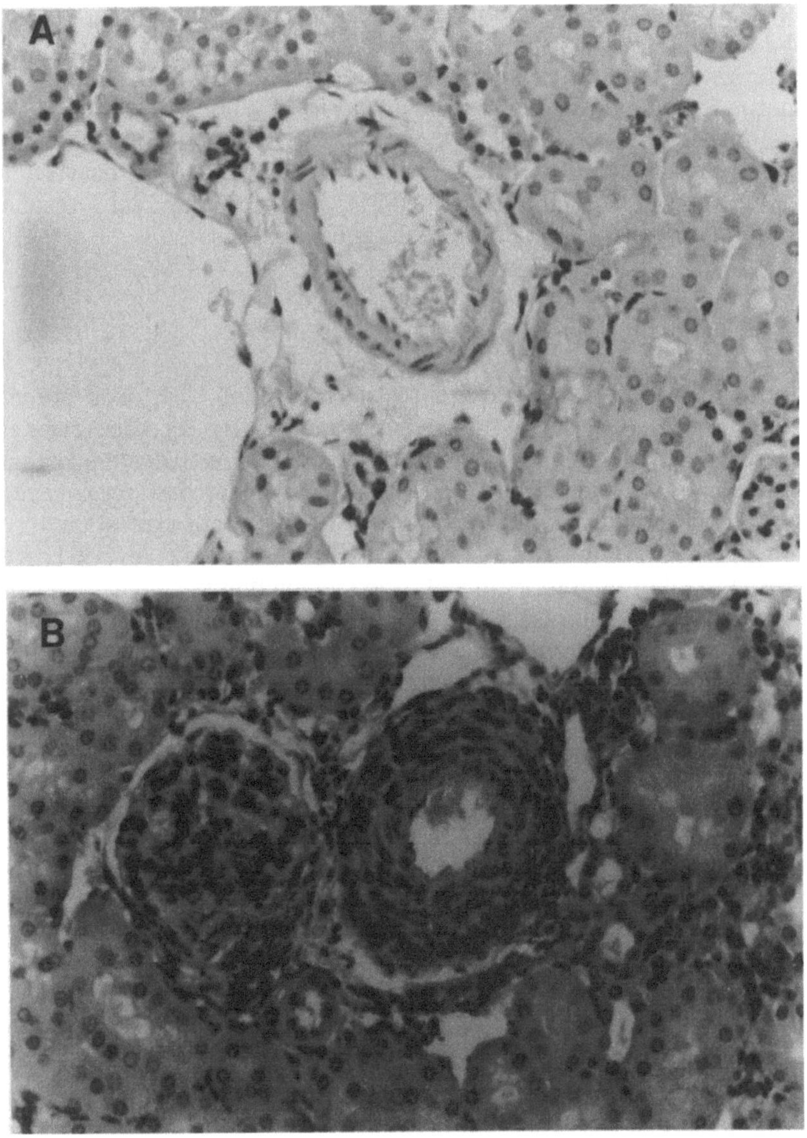

Fig. 3A, B. Renal vascular pathology in adult Ren-Tag mice. Histological sections are shown of kidney tissue from a nontransgenic littermate (**A**) and a Tag 4 transgenic (**B**) stained with hematoxylin and eosin. An arcuate artery is shown in **A**, and an interlobular artery in **B.** Note the thickening of the vessel wall due to hyperplasia of medial smooth muscle cells

correct tissue- and cell-specific expression of this transgene (Sigmund et al. 1990a). Importantly, renin was not observed at nonexpressing sites for renin. Although the relative frequency with which tumors formed at the different sites varied between founder lines, essentially the same spectrum of tumor types was observed for each line.

Efforts have focused on tumors arising in the kidney with the aim of trying to establish a renin-expressing cell line from this source (Sigmund et al. 1990c). The intraparenchymal tumors which developed at 5–8 months of age were unilateral in origin. The general histological appearance of the tumor was consistent with an origin from nephroblastic tissue or from vascular smooth muscle.

Multiple primary tumors and their derivatives propagated in nude mice showed evidence of renin and Tag expresion (Sigmund et al. 1990c). Such expression was maintained over long term passage in cell culture, and a clonal cell line, 4.1, was derived. Figure 4 shows the gross morphology and salient expression characteristics of the 4.1 clonal cell line. The cells uniformly contain nuclei that stain immunocytochemically for Tag, and renin mRNA is readily detectable by in situ hybridization. Quantitative Northern blot hybridization indicates that the renin mRNA abundance level is in the range of thousands of copies of renin mRNA per cell under normal culture growth conditions.

In Fig. 5, a primer extension assay, which can discriminate between the different mouse renin gene transcripts and interspecifically between mouse and human renin gene transcripts, has been used to type the renin mRNA expressed in the 4.1 cell line. As expected, this assay documents that the renin transcript which accumulates is of mouse origin and derived from the *Ren-1*^{c} gene. This is consistent with the genotype of the mouse host of the tumor.

In the course of propagating the 4.1 cell line, occasional fluctuations or drift in the level of renin mRNA were observed to occur. Because these fluctuations

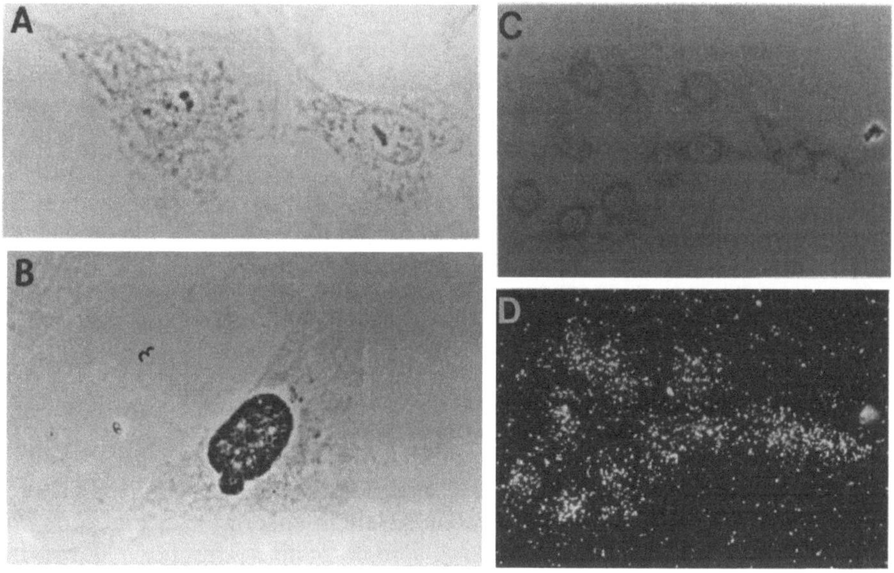

Fig. 4A–D. Cell morphology and expression. Cells were grown on acid-washed glass slides and were subjected to T antigen immunochemistry (**A, B**) or renin in situ hybridization (**C, D**). **A**, STO fibroblasts; **B**, AS 4.1; **C** and **D**, bright field and dark field, respectively, of 4.1 cells

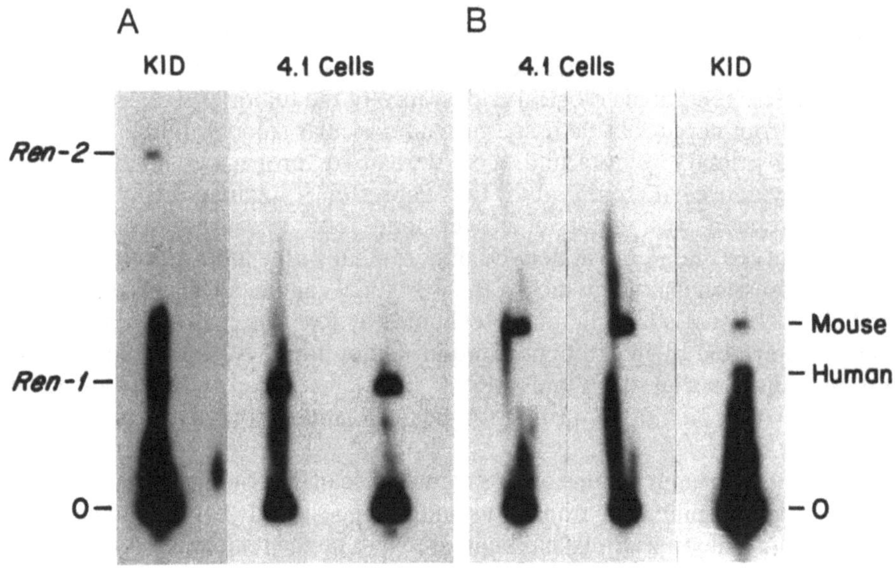

Fig. 5A, B. Renin gene constitution of cell line 4.1. Five μg of RNA from different 4.1 cell lineages were hybridized to a 38 mer, which differentiates *Ren-1* from *Ren-2* (**A**), or a 37 mer, which differentiates mouse renin from human renin transcripts (**B**). Controls: **A** KID, DBA/2Ros kidney RNA (80 μg); **B** KID, control kidney RNA from a transgenic mouse containing the human renin gene and expressing both *Ren-1* and human renin mRNA

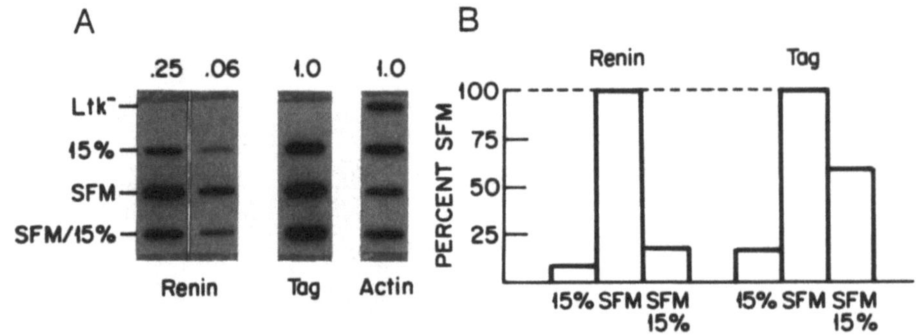

Fig. 6A, B. Serum response: (**A**) slot blot hybridization of RNA isolated from cells grown under differing conditions (the amount of RNA (in μg) is indicated along the top and the probe used along the bottom); (**B**) quantitation of slot blot in **A.** All data are normalized to renin mRNA level of cells grown in serum-free media (SFM) after correcting for RNA loading (actin lane). Cells were grown in the way described in the text

appeared to coincide with changes in the serum lots used in preparing the culture media, a more systematic investigation of the effect of serum concentration on renin mRNA levels was undertaken. A sample of these results is shown in Fig. 6A, B.

The level of renin mRNA is significantly and inversely proportional with the serum concentration in the growth media. Moreover, the kinetics of the

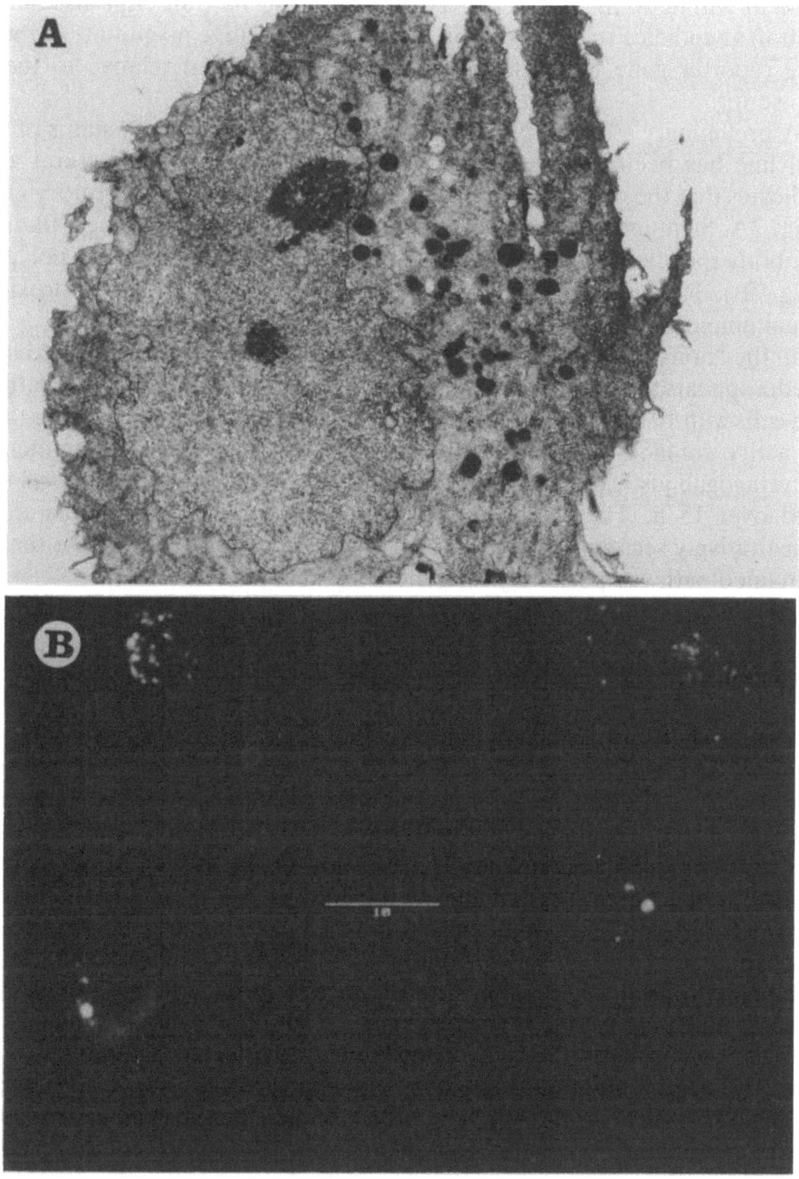

Fig. 7A, B. Demonstration of granules and renin in AS 4.1 cells. **A** An electron micrograph showing the presence of electron-dense granules; original magnification × 5000. **B** An indirect fluorescent confocal micrograph illustrating the intracellular distribution of renin. *Scale bar,* 10 μm

response was relatively rapid, with an almost ten fold increase in mRNA abundance discernible within 24 h of shifting from serum-supplemented media. The reverse shift also exhibited rapid kinetics: upon supplementation of the serum-free media with 15 % FBS, a six-fold decrease in renin mRNA level was evident within 24 h. The *Ren-2*-SV40 Tag transgene transcript also showed a shift in abundance relative to actin control. The relative magnitude of the SV40 Tag reporter gene response was somewhat diminished relative to the renin transcript.

A preliminary assessment of renin activity and the secretory status of the 4.1 cell line has been reported (Sigmund et al. 1990c). Ultrastructural analysis indicates that the cell line contains numerous electron-dense secretory granules (Fig. 7A, Sigmund et al. 1990c). Confocal microscopy employing a fluorescent antibody specific for renin suggests that the granules contain renin polypeptide (Fig. 7B). Renin activity assays suggest that the cells contain approximately equal amounts of renin and prorenin (Sigmund et al. 1990c).

In the "nonstimulated" state, 95 % of the renin polypeptide released into the media appears to be in the form of prorenin (Sigmund et al. 1990c). Treatment of the cells with 10 m*M* 8-Br-cAMP for 4 or 15 h has been reported to raise the level of active renin in the media 2.3× and 4.6×, respectively. Treatment with secretagogue also appeared to raise the renin mRNA abundance level several fold over 15 h. These results are consistent with the notion that the cells constitutively secrete prorenin but may be induced to secrete active renin by a regulated pathway positively influenced by cyclic AMP levels.

Discussion

The studies described clearly support the assertion that 4.6 kb of *Ren-2* 5′ flanking sequence is sufficient to confer a correct tissue- and cell-specific expression profile on a reporter gene in transgenic mice. Moreover, the pattern observed is ontogenetically precise through early adulthood. This suggests that relevant *cis-acting* regulatory sequences lie within this DNA segment. The studies refine our understanding of the localization of important regulatory elements relative to previous studies, in which large natural genomic segments containing both introns and considerable 3′ flanking sequences were shown to exhibit appropriate expression. In addition these results, in the context of other reports which indicate that the proximal 2.5 kb of 5′ flanking sequence is not sufficient to properly direct expression of a reporter gene (Sola et al. 1989), strongly suggest important regulatory sequence lie distal from 2.5 kb but closer than 4.6 kb relative to the CAP site of the *Ren-2* gene. Such an interpretation is gratifyingly compatible with recent interspecific comparisons of the structure of renin genes. The mouse genes, when compared to either the rat or human renin genes, exhibit evidence of DNA insertion immediately proximal to the CAP site (Soubrier et al. 1986; Abel and Gross 1990). Such insertions may have pushed important DNA regulatory elements in the mouse more distally.

In addition to providing an exquisitely sensitive reporter function, T antigen has oncogenic potential. The ramifications of this potential are manifest in older animals in the diffuse renal vascular pathology, which is observed, as well as overt tumorigenesis, at renin expression sites.

As described, adult transgenic animals exhibit hyperplasia of T-antigen containing SMC throughout the renal arterial tree. This is not a normal site of expression in adult animals, although expression is seen at such sites during fetal life and can be induced under certain pathophysiological conditions in adults (Jones et al. 1990; Lindop and Lever 1986; Minuth et al. 1981; Richoux et al. 1987; Gomez et al. 1986, 1988a, b, 1989). The body of experimental data suggests that the observed expression of Tag reflects reactivation of Tag expression at these sites rather than persistence of expression initiated during fetal life. This suggests that SMC, even in the major intra-renal arteries, retains the *potential* to express renin upon appropriate pathophysiologcial stimulation. The precise mechanism underlying the recruitment of SMC back into the renin-expressing population is unclear at present, but it may reflect the progressive vascular hypertrophy and presumed attendant ischemia.

The diffuse renal SMC hyperplasia is accompanied by a variable incidence of intimal fibrosis. The evidence suggests that these lesions arise secondarily. Remarkably, despite the pronounced renal vascular pathology, these transgenic animals appear normotensive. In collaboration with V. Dzau's group and as described in Sigmund et al. (1991) and Jacob et al. (1991), a preliminary analysis of the physiological status of these animals has been performed. Adult animals exhibit marked suppression of renal renin mRNA, renal renin content and plasma renin concentration. Additionally they show extracellular volume expansion without evidence of renal failure. Therefore, these animals may provide an interesting and novel animal model with which to investigate the physiological correlates of primary vascular hyperplasia and the control of blood pressure in a low renin state.

Ultimately, T antigen expression predisposes these animals to overt tumorigenesis. The spectrum of tissues in which tumorigenesis is observed to occur, in addition to direct analysis of transgene expression, serves to support the contention that the 4.6 kb upstream region is directing the oncogenes' expression with fidelity (Sigmund et al. 1990a). The tumors which develop appear focal in origin, and they form at frequencies which support the notion that tumorigenesis is a stochastic and multistep process.

The spectrum of tumors which form may ultimately provide the basis for developing a library of cell lines with which to study aspects of renin expression at different organ sites in detail. For the present, we have focused on establishing and studying cell lines derived from kidney tumors. These tumors are of apparent nephroblastic origin, although the precise cell type giving rise to them may be difficult to determine.

At present a clonal cell line, herein described as 4.1, has been carried in continuous culture for six months (approximately 35 passages). The doubling time in DMEM/15 % FBS media is on the order of 24 h. This cell line clearly has a number of characteristics which are of interest vis-a-vis its use as a model cell for studying the regulation of renin expression and secretion. The 4.1 cell line exhibits high abundance of the cognate *Ren-1*^{c} mRNA. The level of renin mRNA appears to be highly sensitive to the composition of the growth media, particularly the concentration and type of serum component. Indeed a marked inverse correlation between the concentration of serum and the level of renin mRNA is observed. The fluctuations in renin mRNA level are characterized by rapid kinetics. It will be of interest to establish the relative contributions of transcription and turnover parameters to this effect. It will also be important to establish what aspect or component of serum is responsible for the observed change and whether other genes are similarly affected. (There appear to be clear parallels in this regard with serum effects on primary cultures of SMC; Chamley-Campbell et al. 1981).

The cell line secretes prorenin into the media constitutively but accumulates both renin and prorenin in apparently equal abundance intracellulary. Ultrastructural studies suggest that renin is localized in dense granules, and preliminary studies suggest that the cell retains components of a regulated secretory pathway. It will be of interest to subject these cells to more detailed scrutiny to ascertain what differentiated features have been retained in culture and how extensively they mimic the renin secretory cell in situ. Such characterizations will necessarily include assessment of membrane receptor populations, operant signal transduction pathways and detailed examination of secretory pathways.

Clearly, the cells also provide an important source of trans-acting factors that interact with renin regulatory sequences.

Acknowledgment. We would like to thank Marcia Held for excellent secretarial assistance.

References

Abel KJ, Gross KW (1990) Physical characterization of genetic rearrangements at the mouse renin loci. Genetics 124: 937–947

Ausubel FM, Crent R, Kingston RE, Moore DD, Smith JA, Seidman JG, Struhl K (1987) Current protocols in molecular biology. Wiley, New York

Chamley-Campbell JH, Campbell GR, Ross R (1981) Phenotype-dependent response of cultured aortic smooth muscle to serum mitogens. J Cell Biol 89: 379–383

Ekker M, Tronik D, Rougeon F (1989) Extra-renal transcription of the renin genes in multiple tissues of mice and rats. Proc Natl Acad Sci USA: 86: 5155–5158

Fabian J, Field LJ, McGowan RA, Mullins JJ, Sigmund CD, Gross KW (1989) Allele-specific expression of the murine Ren-1 genes. J Biol Chem 264: 17589–17594

Field LJ, Gross KW (1985) Ren-1 and Ren-2 loci are expressed in mouse kidney. Proc Natl Acad Sci USA 82: 6196–6200

Field LJ, McGowan RA, Dickenson DP, Gross KW (1984) Tissue and gene specificity of mouse renin expression. Hypertension 6: 597–603

Gomez RA, Chevalier RL, Sturgill BC, Johns DW, Peach MJ, Carey RM (1986) Maturation of the intrarenal renin distribution in Wistar-Kyoto rats. J Hypertens 4: S31–S33

Gomez RA, Lynch KR, Chevalier RL, Everett A, Johns DW, Wilfong N, Peach MJ, Carey RM (1988a) Renin and angiotensinogen gene expression and intrarenal renin distribution during ACE inhibition. Am J Physiol 254: F900–F906

Gomez RA, Lynch KR, Chevalier RL, Wilfong N, Everett A, Carey RM, Peach MJ (1988b) Renin and angiotensinogen gene expression in maturing rat kidney. Am J Physiol 254: F582–F587

Gomez RA, Lynch KR, Sturgill BC, Elwood JP, Chevalier RL, Carey RM, Peach MJ (1989) Distribution of renin mRNA and its protein in the developing kidney. Am J Physiol 257: F850–F858

Jacob H, Sigmund CD, Shockley TR, Gross KW, Dzau VJ (1991) Renin promoter-SV40 T-antigen transgenic mouse: a model of primary renal vascular hyperplasia. Hypertension 17: 1167–1172

Jones CA, Sigmund CD, McGowan R, Kane-Haas C, Gross KW (1990) Temporal and spatial expression of the murine renin genes during fetal development. Mol Endocrinol 4: 375–383
Lindop GEM, Lever AF (1986) Anatomy of the renin-angiotensin system in the normal and pathologic kidney. Histology 76: 61–69
Lynn DA, Angerer M, Bruskin AM, Klein WH, Angerer RC (1983) Localization of a family of mRNAs in a single cell type and its precursors in sea urchin embryos. Proc Natl Acad Sci USA 80: 2656–2660
Miller CCJ, Canter AT, Brooks JF, Lovell-Badge RH, Brammer WJ (1989) Differential extra-renal expression of the mouse renin genes. Nucleic Acids Res 17: 3117–3128
Minuth M, Hackenthal E, Poulsen K, Rix E, Taugner R (1981) Renin immunocytochemistry of the differentiating juxtaglomerular apparatus. Anat Embryol 162: 173–181
Mullins JJ, Sigmund CD, Kane-Haas C, Wu C, Pacholec F, Zeng Q, Gross KW (1988) Studies on the regulation of renin genes using transgenic mice. Clin Exp Hypertens [A] 10: 1157–1167
Mullins JJ, Sigmund CD, Kane-Haas C, Gross KW (1989) Expression of the DBA/2 Ren-2 gene in the adrenal gland of transgenic mice. EMBO J 8: 4065–4072
Naruse M, Susson CR, Naruse K, Jackson RV, Inagami T (1983) Renin exists in human adrenal tissue. J Clin Endocrinol Metab 57: 462–487
Piccini N, Knopf JL, Gross KW (1982) A DNA polymorphism, consistent with gene duplication correlates with high renin levels in the mouse submaxillary gland. Cell 30: 205–213
Richoux JP, Amsaguine S, Grignon G, Bouhnik J, Menard J, Corvol P (1987) Earliest renin containing cell differentiation during ontogenesis in the rat. Histochem J 88: 41–46
Sealey JE, Glorioso N, Itskovitz J, Laragh JH (1986) Prorenin as a reproductive hormone: new form of the renin system. Am J Med 81: 1041–1046
Sigmund CD, Gross KW (1990) Differential expression of the murine and rat renin genes in peripheral subcutaneous tissue. Biochem Biophys Res Commun 173: 218–223
Sigmund CD, Jones CA, Fabian JR, Mullins JJ, Gross KW (1990a) Tissue and cell-specific expression of a renin promotor-T antigen reporter gene construct in transgenic mice. Biochem Biophys Res Com 170: 344–350
Sigmund CD, Jones CA, Kim U, Mullins JJ, Gross KW (1990b) Expression of the murine renin genes in subcutaneous connective tisue. Proc Natl Acad Sci USA 87: 7993–7997
Sigmund CD, Okuyama K, Ingelfinger J, Jones CA, Mullins JJ, Kim U, Kane-Haas C, Wu C, Kenney L, Rustum Y, Dzau V, Gross KW (1990c) Isolation and characterization of renin expressing cell lines from transgenic mice containing a renin promoter viral oncogene fusion construct. J Biol Chem 265: 19916–19922
Sigmund CD, Jones CA, Jacob H, Ingelfinger J, Kim U, Gamble D, Dzau VJ, Gross KW (1991) Pathophysiology of vascular smooth muscle in renin promoter-T antigen transgenic mice. Am J Physiol 260: F249–F257
Sola C, Tronik D, Dreyfus M, Babinet C, Rougeon F (1989) Renin-promoter SV40 large T antigen transgenes induce tumors irrespective of normal cellular expression of renin genes. Oncogene Res 5: 149–153
Soubrier F, Panthier JJ, Houot AM, Rougeon F, Corvol P (1986) Segmental homology between the promoter region of the human renin gene and the mouse renl and ren2 promoter regions. Gene 41: 85–92
Tronik D, Dreyfus M, Babinet C, Rougeon F (1987) Regulated expression of the Ren-2 gene in transgenic mice derived from parental strains carrying only the Ren-1 gene. EMBO J 6: 983–987

Tissue Distribution of Ren-2 Expression in Transgenic Rats

J. Peters, M. Bader, D. Ganten, and *J. Mullins*

Summary

Transgenic rats harboring and expressing the mouse *Ren-2* gene were constructed by pronuclear microinjection of fertilised oocytes. Two strains of hypertensive transgenic rats have been successfully established, one of which, TGR(mRen2)27, has been the subject of detailed investigation. These rats have been shown to exhibit extreme hypertension but have a suppressed circulating renin angiotensin system. The focus of the present study was to define those tissues which express the transgene and to determine which tissues may be involved in the pathogenesis of the hypertensive phenotype observed. From a survey of a wide range of tissues, the most active site of transgene expression was found to be the adrenal gland. However several other tissues were shown to express the transgene at significant levels, including the vasculature and segments of the gastrointestinal tract. These data will, in part, aid in the understanding of this new model of hypertension.

Introduction

Primary hypertension is the main risk factor for cardiovascular morbidity and mortality, but the complex interactions leading to chronically elevated blood pressure remain poorly understood. The renin angiotensin system (RAS) is one of the key modulators of cardiovascular homeostasis. The aspartyl protease renin is synthesized intracellularly as preprorenin, cleaved to release prorenin and finally activated to form the active enzyme. This activation process takes place in the storage granules of the juxtaglomerular cells, from which renin is secreted in a regulated manner; however, the proenzyme may be secreted by a constitutive pathway. Renin acts on its only known substrate angiotensinogen to generate angiotensin I, which is in turn cleaved by dipeptidyl carboxypeptidase (converting enzyme) to form the vasopressor peptide angiotensin II. Angiotensin II has a direct vasoconstrictor action and is – due to its own action on kidney tubules as well as stimulating aldosterone secretion – an important regulator of intracellular and extracellular water and electrolyte balance. Although the primary source of circulating renin is the juxtaglomerular apparatus of the kidney, recent data strongly support the existence of local renin-angiotensin

K. Berg, V. Bulyzhenkov, Y. Christen, P. Corvol (Eds.)
Genetic Approaches to Coronary Heart Disease
and Hypertension

systems which may act in a paracrine or autocrine manner (Ganten et al. 1984; Dzau 1988; Lindpaintner et al. 1989). Extra-renal sources of renin have been described, including the submandibular gland, ovary, testis, adrenal gland, brain and heart (Deschepper et al. 1986; Samani et al. 1988; Paul et al. 1988; Ekker et al. 1989), and the physiological significance of expression in these tissues is undergoing investigation.

Recently, the molecular biology of the RAS has been well studied and the genes involved have been cloned, and characterised both in vitro and in vivo (Miyazaki et al. 1984; Fukamizu et al. 1990). Most mammals possess a single renin structural gene but certain strains of inbred mice and some wild mouse species carry a duplicated renin gene, the *Ren-2* gene (Mullins et al. 1982). This gene is closely linked to the *Ren-1* gene (Abel and Gross 1990) but has a distinct profile of tissue-specific expression despite a high degree of sequence homology. Although the physiological role of the Ren-2-encoded enzyme is not fully understood, purified protein injected in vivo has been shown to have biological activity. The *Ren-2* gene has been well characterised at the molecular level both in vitro and in vivo using transgenic mice (Tronik et al. 1987; Mullins et al. 1989). These studies demonstrated that the sequences necessary for appropriate tissue-specific expression were present within or close to the renin structural gene (Mullins et al. 1988, 1989; Makamura et al. 1989). The gene was expressed qualitatively in the expected range of tissues, and with appropriate hormonal control in the submandibular gland. Studies using a reporter construct have further defined sequences important for tissue specific expression of the *Ren-2* gene (Sigmund et al. 1990). In addition to these studies it was of interest to begin to use this information to develop animals with defined alterations in the renin-angiotensin system. This report summarises data on the expression of the *Ren-2* gene in tissues of transgenic rats harboring the *Ren-2* transgene.

Transgenic Rats

To enhance the possibilities afforded by transgenic technology we recently modified the techniques to efficiently generate transgenic rats (Mullins et al. 1990). The relative merits of using the rat compared to the mouse are mainly due to its physical size, which is conducive to certain cardiovascular, physiological and pharmacological studies otherwise difficult or impossible in the mouse. This had to be balanced against cost of maintenance, and paucity of genetic characterisation in the rat (e.g., chromosomal mapping of genetic loci, availability of congenic strains) compared to the extensive information available for the mouse. Of greater significance was the lack of detailed knowledge regarding ovulation in the rat. However, the possibility for precise manipulation of genes involved in blood pressure homeostasis in the rat was felt to be an important and achievable goal.

To improve ovulation of the donor animals, and thus reduce the number of animals required, immature female rats (4 weeks old) were superovulated according to the procedure of Armstrong and Opavsky (1988). By administering a highly purified preparation of follicle-stimulating hormone (isolated from pig pituitary) with an Alzet mini-osmotic pump, the problem of hormone instability was overcome, and we routinely obtained approximately 80 fertilized eggs per superovulated female. This represents a six- to eight-fold increase over normal ovulation. Minor modifications to the microinjection procedure had to be made due to the different physiological and physical properties of the eggs. Together with similar modifications to the surgical procedures for reimplantation of the microinjected eggs, we achieved a high rate of pregnancy (greather than 70 %) in recipient pseudopregnant Sprague Dawley females.

To establish the technique we first introduced the mouse *Ren-2* gene into the rat germline (Mullins et al. 1990). The TGR(mRen2) rats generated by these methods integrated the entire DBA/2 mouse *Ren-2* gene including 5.3 kb of 5′ flanking DNA and 9.3 kb of 3′ flanking DNA. These rats developed fulminant hypertension beginning at five weeks of age and reaching approximately 220 mm Hg at 8–10 weeks of age versus 120 mm Hg in the normotensive controls. Homozygous TGR(mRen2) transgenic rats exhibit even higher blood pressure (unpublished data). The evidence indicates that this phenotype is not due to insertional mutagenesis, since four independent hypertensive transgenic animals were generated, each of which integrated the transgene at a different chromosomal position. The hypertension is therefore due directly or indirectly to the expression of the transgene. Studies on one line of transgenic rats, TGR(mRen2)27, revealed that the hypertension could be controlled by converting enzyme inhibition with a low dose of captopril. These data have been confirmed in subsequent studies in our laboratory and extended to include other converting enzyme inhibitors (data in preparation). The elevated blood pressure in these rats was accompanied by a suppression of kidney renin (which was presumably due to normal feedback mechanisms), and also a suppressed plasma renin-angiotensin system. However, the plasma prorenin levels were highly elevated, suggesting an extra-renal source of prorenin.

Tissue Survey

To determine whether the transgene is still under specific regulatory control and maintains tissue-specific expression and to select candidate tissues expressing or affected by the transgene, which may be involved in the mechanism of the hypertension, we screened tissues of the transgenic rats for the presence of renin transcripts. In total 37 tissues from male and female transgene-positive rats and their corresponding transgene-negative littermates were analysed by Northern blotting (Fig. 1) and RNAse protection techniques (Fig. 2). One of the primary tissues found to be expressing the transgene was the adrenal gland where *Ren-2* specific transcripts were abundant. We estimate that the level of expression of the *Ren-2* transgene in the transgenic adrenal gland was at least 2500-fold greater than that of the endogenous rat renin gene. The level of expression of the transgene was 10 to 500 times higher in the adrenal gland than in any other tissue assayed. Although transgene expression was evident in the kidney the expression was found to be very low, reflecting suppression of the kidney RAS. One tissue of particular interest is the submandibular gland which, in the DBA/2

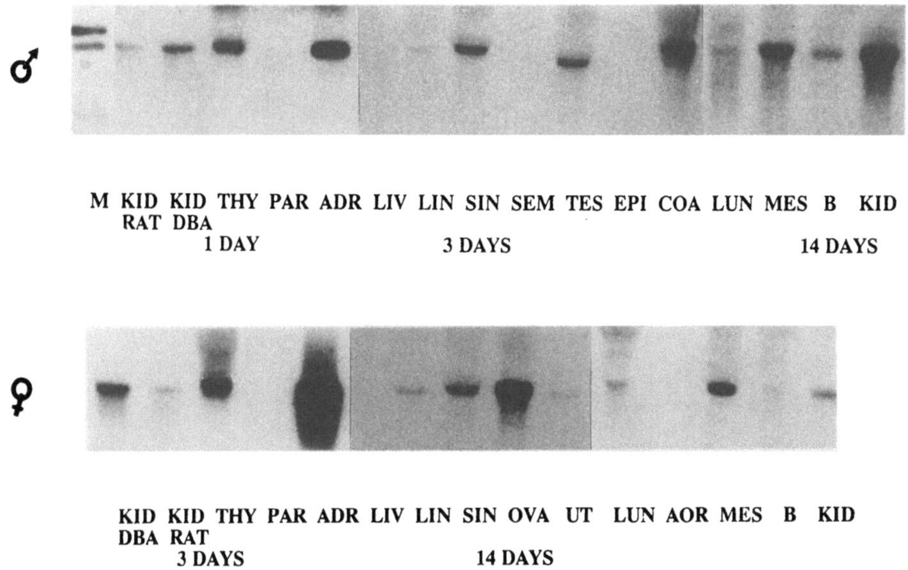

Fig. 1. Northern Blot Analysis. Northern blot analysis was performed on total RNA samples from tissues of transgenic animals as follows: In each lane 50 μg RNA was loaded, with the exceptions of adrenal gland (5 μg) and DBA/2 kidney (20 μg). Total RNA preparations were denatured in 17 m*M* NaH_2PO_4 pH 7.0, 1 *M* glyoxal, and 9 *M* DMSO at 50 °C for 50 min, cooled in ice water, and electrophoresed on a 1.25 % agarose gel (loading buffer was 1/5 volume of: 50 % glycerol, 0.25 % bromophenol blue, 0.25 % xylene cyanol and 10 m*M* Tris-HCl pH 7.0) under circulation in 10 m*M* sodium phosfate buffer pH 6.8. Electrophoresed samples were transferred to nytran nylon membrane (Scheicher and Schull) using a vacuum blot system and baked at 80 °C for 2 h. Filters were prehybridized at 62 °C in 50 % formamide, 5 × SSC, 1 % SDS, 5 × Denharts solution and 200 μg/ml denatured salmon sperm DNA, followed by overnight hybridization at 62 °C in a similar solution containing 10 % dextran sulfate and a ^{32}P UTP-labelled cRNA probe prepared from the Ren-2 cDNA clone pSLM (Mullins et al. 1989) at 5×10^6 cpm/ml. Filters were then washed for 15 min with 2 × SSC, 0.1 % SDS at room temperature followed by 2 h washing with 0.1 × SSC, 0.1 % SDS at 65 °C, and then placed to film.

mouse, expresses the *Ren-2* gene at a high level. Expression of the transgene was not detected in the submandibular gland of the transgenic rats with the assays used in the present study, suggesting the absence of appropriate trans-acting factors within the submandibular gland of the rat. With the exceptions of the submandibular gland, the tissue distribution of *Ren-2* expression is qualitatively similar to that found in the mouse. *Ren-2* mRNA was detected in adrenal, kidney, brain (cortex/septum), pituitary, aorta, mesenteric vessals, lung, testis, coagulation gland, ovary, uterus and thymus. Weak expression was also observed in the thyroid gland and the eye. Expression was not detected in liver, parotid gland, lacrimal gland, heart, cerebellum, medula oblongata, hypothalamus, submandibular gland, muscle, epididymis, seminal vesicle or spleen.

A surprising site in which expression was detected was the gastrointestinal tract, where expression was localised to the stomach and small intestine. The stomach and small intestine were subsequently shown to be bona fide sites of renin expression in the mouse (data not shown), and therefore the tissue

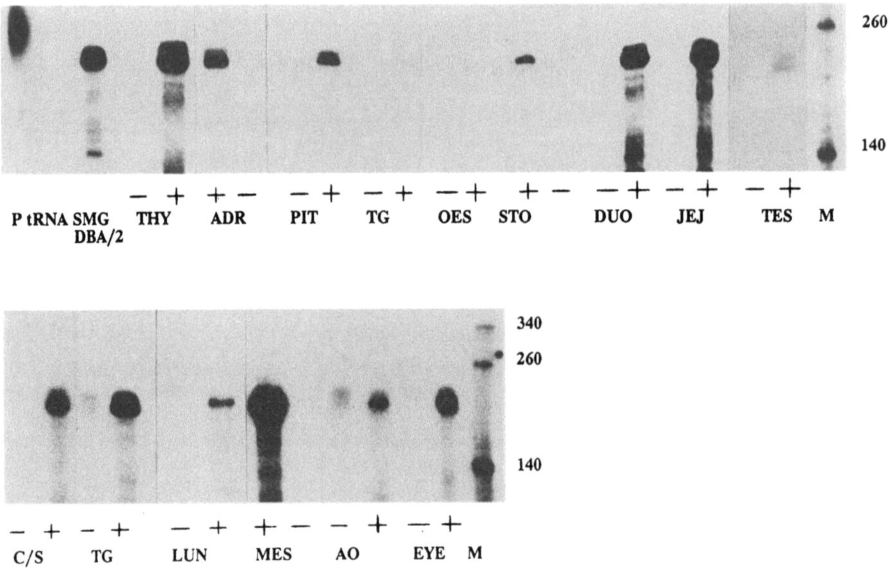

Fig. 2. The Ren-2 mRNA-specific RNAse protection assay. The assay was performed using total RNA isolated from the respective tisues of transgenic and nontransgenic male rats. ^{32}P-labelled RNA transcripts were prepared by transcription of a 244-nucleotide antisense RNA from the plasmid pSLM (Mullins et al. 1989) using SP6 RNA polymerase. This transcript comprised 224 nucleotides of Ren-2 antisense RNA and 20 nucleotides of vector-encoded sequence. Samples were dissolved in 30 µl 80 % formamide, containing 40 m*M* PIPES, 400 m*M* NaCl, 1 mM EDTA and 200 000 cpm of the gel-purified transcript, denatured at 100 °C for 1 min and incubated at 45 °C for 20 h. RNase digestion was performed in 300 µl buffer containing 40 µg/ml RNase A (Sigma) and 2 µg/ml RNase T1 (Calbiochem) for 45 min at 37 °C. After digestion with proteinase K, samples were electrophoresed on denaturing 5 % polyacrylamide gels.
Fifty µg of total RNA was used for each sample except for DBA/2 SMG (0.25 µg) and adrenal gland (2.5 µg). Samples from the transgene-positive rat are indicated by +, and those from transgene-negative littermate are indicated by −. Size markers (M) are shown to the right of the figure and are given in bases. *P,* undigested RNA transcript used in the protection assay; *tRNA,* RNAse-free *E. coli* tRNA control; *SMG DBA/2,* male DBA/2 mouse submandibular gland; *THY,* thymus; *ADR,* adrenal gland; *PIT,* pituitary gland; *TG,* thyroid gland; *OES,* esophagus; *STO,* stomach; *DUO,* duodenum; *JEJ,* jejunum; *TES,* testis; *C/S,* cortex/septum; *LUN,* lung; *MES,* mesenteric vessels; *AO,* aorta;, *EYE,* eye. The *upper panel* was exposed to film for 24 h; the *lower panel* was exposed to film for 14 days except the *C/S lane* (21 days). All negative control lanes were devoid of signal even after a 14-day exposure

distribution within the gastrointestinal tract was consistent with appropriate expression of the mouse gene. Using a *Ren-2*-derived hybridization probe, endogenous rat renin gene expression in the control transgene-negative animals was only detectable in the kidney. However, a rat renin mRNA-specific RNAse protection assay was not used in the present study. In conclusion, the expression pattern of the DBA/2 *Ren-2* gene strongly suggests that the tissue specificity of the *Ren-2* gene is in large part maintained. There are differences within specific tissues, however, which demonstrate that there is an interaction between the cis regulation of the transgene sequences and the trans-regulation of available transcription factors (data in preparation).

Conclusion

We have further defined the tissue specific expression pattern of the *Ren-2* transgene in the TGR(mRen-2)27 animals. Following a detailed tissue survey, the available data confirm that the adrenal gland is the major site of *Ren-2* expression in the hypertensive rats, and this tissue is the focus of present experiments to determine any causal relationship to the hypertensive phenotype. This model may well represent a model for low plasma renin hypertension with a genetically defined abnormally of the renin-angiotensin system.

The extension of transgenic technology to include the rat is an important development, both for hypertension research and for wider research applications. Recently, a second example of transgenic rats was published, where the human B27 and β_2-microglobulin genes were introduced into the rat. These caused an inflammatory disease similar to the human B27-related arthritic syndromes (Hammer et al. 1990). One striking feature, both of this study and the present study, is that no equivalent model was generated when the respective genes were introduced into the mouse genome (Taruog et al. 1988; Krimpenfort et al. 1987; Mullins et al. 1988). It would therefore appear that choice of species is an important consideration in the design of animal models. Studies of the expression of the *Ren-2* gene during ontogeny and pharmacological treatment should help elucidate the mechanism for the hypertension in these transgenic rats.

Acknowledgements. We thank Gesa Wernicke for excellent technical assistance and Dr. L. J. Mullins for contributions in preparing the manuscript.

References

Abel K, Gross KW (1990) Physical characterisation of genetic rearrangements at the mouse renin loci. Genetics 124: 937–947

Armstrong DT, Opavsky MA (1988) Superovulation of immature rats by continuous infusion of FSH. Biol Reprod 39: 511–518

Deschepper CF, Mellon SH, Cumin F, Baxter JD, Ganong WF (1986) Analysis of immunocytochemistry and in situ hybridization of renin and its mRNA in kidney, testis, adrenal and pituitary of the rat. Proc Natl Acad Sci USA 83: 7552–7556

Dzau VJ (1988) Circulating versus local renin angiotensin system in cardiovascular homeostasis. Circulation [suppl 1] I/4–I/13

Ekker M, Tronik D, Rougeon F (1989) Extra-renal transcription of the renin genes in multiple tissues of mice and rats. Proc Natl Acad Sci USA 86: 5155–5158

Fukamizu A, Takahashi S, Seo M.S, Tada M, Tanimoto K, Uehara S, Murakami K. (1990) Structure and expression of the human angiotensinogen gene. J. Biol. Chem. 265: 7576–7582

Ganten D, Lang RE, Lehmann E, Unger T (1984) Brain angiotensin: On the way to becoming a well studied neuropeptide system. Biochem Pharmacol 33: 3523–3528

Hammer RE, Maika SD, Richardson JA, Tang JP, Taurog JD (1990) Spontaneous inflammatory disease in transgenic rats expressing HLA-B27 and human β_2 microglobulin: an animal model of HLA-B27-asociated human disorders. Cell 63: 1099–1112

Krimpenfort P, Rudenko G, Hochstenbach F, Gussow D, Berns A, Ploegh H (1987) Crosses of two independently derived transgenic mice demonstrate functional complementation of the genes encoding heavy (HLA-B27) and light (β_2-microglobulin) chains of HLA class 1 antigens. EMBO J 6: 1673–1676

Lindpaintner K, Jin M, Wilhelm M, Toth M, Ganten D (1989) Aspects of the molecular biology and biochemistry of the cardiac renin-angiotensin system. Br J Clin Pharmacol 27: 159S–165S

Makamura N, Burt DW, Paul M, Dzau VJ (1989) Negative control elements and cAMP responsive sequences in the tissue-specific expression of mouse renin genes. Proc Natl Acad Sci USA 86: 56–59

Miyazaki H, Fukamizu A, Hirose S, Hayashi T, Hori H, Ohkubo H, Nakanishi S, Murakami K (1984) Structure of the human renin gene. Proc Natl Acad Sci USA 81: 5999–6003

Mullins JJ, Burt DW, Windass JD, McTurk P, George H, Brammar WJ (1982) Molecular cloning of two distinct renin genes from the DBA/2 mouse. EMBO J 1: 1461–1466

Mullins JJ, Sigmund CD, Kane-Haas C, Wu C, Pacholec F, Zeng Q, Gross KW (1988) Studies on the regulation of renin genes using transgenic mice. Clin Exp Hypertens [A] 10 (6): 1157–1167

Mulins JJ, Sigmund CD, Kane-Haas C, Gross KW (1989) Expression of the DBA/2J Ren-2 gene in the adrenal gland of transgenic mice. EMBO J 8: 4065–4072

Mullins JJ, Peters J, Ganten D (1990) Fulminant hypertension in transgenic rats harbouring the mouse Ren-2 gene. Nature 344: 541–544

Paul M, Wagner D, Metzger R, Ganten D, Lang RE, Suzuki F, Murakami K, Burbach JHP, Ludwig G (1988) Quantification of renin mRNA in various mouse tissues by a novel solution hybridization assay. J Hypertens 6: 247–252

Samani J, Swales D, Brammar WJ (1988) Expression of the renin gene in extra-renal tissue of rat. Biochem J 253: 907–910

Sigmund CD, Jones CA, Fabian JR, Mullins JJ, Gross KW (1990) Tissue and cell specific expression of a renin promoter-reporter gene construct in transgenic mice. Biochem Biophys Res Commun 170: 344–350

Taurog JD, Lowen L, Forman J, Hammer RE (1988) HLA-B27 in inbred and non-inbred transgenic mice. Cell surface expression and recognition as an alloantigen in the absence of β_2-microglobulin. J Immunol 141: 4020–4023

Tronik D, Dreyfus M, Babinet C, Rougeon F (1987) Regulated expression of the Ren-2 gene in transgenic mice derived from parental strains carrying only the Ren-1 gene. EMBO J 6: 983–987

Prediction of Hypertension and Stroke for Controlling Cardiovascular Diseases in Genetic Rat Models

Y. Yamori, Y. Nara, T. Nabika, K. Ikeda, and *K. Matsumoto*

Summary

The development of genetic models for research on hypertension, stroke and atherosclerosis, namely spontaneously hypertensive rats (SHR), stroke-prone SHR (SHRSP) and arteriolipidosis-prone rats (ALR), has contributed not only to the elucidation of the pathogenesis of hypertension-related cardiovascular diseases (CVD) but also to their prediction and prevention. Since both genetic and environmental factors are involved in the pathogenesis of CVD, the detection of genes or genetic factors involved in the pathogenesis and the control of environmental factors, such as dietary improvement, are useful as predictive and preventive measures against CVD. Early rise in blood pressure, sympathetic over-responsiveness, early development of cardiovascular hypertrophy, increased salt sensitivity and membrane or transport abnormalities detected in vascular smooth muscle cells (VSMC) and blood cells from SHR and SHRSP, which appear to be related to the pathogenesis of hypertension, are now regarded as predictors for hypertension possibly applicable to human hypertension, and have been partly examined, together with possible genetic markers and some restriction fragment length polymorphisms in the F_2 generations obtained from SHRSP and normotensive Wistar-Kyoto, for their cosegregation with blood pressure. Genetic pathogenic mechanisms of stroke in SHRSP are hypertension-induced VSMC degeneration and necrosis of intracerebral arteries which start at the outer layer of the media and are accelerated by macrophage infiltration. Predictors of stroke related to these genetic pathogenic mechanisms are reduction of regional cerebral blood flow, hypoxic vulnerability and membrane abnormalities of VSMC, and the acceleration of lipid incorporation into VSMC observed in SHRSP and ALR may be utilized as the cellular predictor of atherosclerotic diseases. On the other hand, some nutritional factors affecting cellular, neural and vascular mechanisms of hypertension and related CVD have been proven to be effective for the prevention of CVD in these genetic models.

K. Berg, V. Bulyzhenkov, Y. Christen, P. Corvol (Eds.)
Genetic Approaches to Coronary Heart Disease and Hypertension

Introduction

Prediction of diseases is essential for primary prevention of diseases (Yamori 1989a). Detection of genetic factors or genes involved in diseases is the most reliable way of predicting diseases. Our studies on rat models for hypertension and stroke, spontaneously hypertensive rats (SHR; Okamoto and Aoki 1963; Yamori 1984a) and stroke-prone SHR (SHRSP; Yamori et al. 1974; Yamori 1984b) have shown that hypertension and stroke can be predicted by detecting initial genetic mechanisms of the development of these diseases.

In this article, the following issues will be covered:
1. the background of SHR models for hypertension and stroke;
2. problems related to genetic variability of these models and control normotensive Wistar-Kyoto rats (WKY);
3. our recent analyses on DNA polymorphisms in search of genes possibly related to hypertension and stroke; and
4. preliminary findings of ongoing studies on the cosegregation in F_2 generations obtained by cross-breeding between SHRSP and WKY.

Finally the pathogenesis and prediction of diseases, that is, the detection of disposition to hypertension, atherosclerosis and stroke will be elucidated, emphysizing their importance in the primary prevention of these cardiovascular diseases (CVD), hopefully contributing to healthy longevity in mankind.

Genetic Background of SHR Models for Hypertension and Stroke

SHR and SHRSP are now widely distributed in the world and regarded as the most useful genetic models for hypertension and stroke in man (Yamori 1983). These models were derived from WKY, which were introduced to Kyoto University from the Wistar Institute through Tokyo and Hokkaido Universities in Japan (Matsumoto et al. 1991). Since genetic analysis of hypertension-related factors were started in these models and control WKY by several authors, including ourselves (Yamori 1983, 1989a; Matsumoto et al. 1991; Kurtz et al. 1989), genetic heterogeneities in these strains have been noted. However these heterogeneities are predictable once one sees the pedigree indicating how SHR, SHRSP and WKY were established and donated directly or indirectly from our original strains maintained at Kyoto University at that time (Fig. 1). Three SHRSP strains and four stroke-resistant SHR (SHRSR) now in Izumo (SHRSP/Izm, SHRSR/Izm) were derived from A, B and C sublines of SHR separated from each other at F13 generation. Arteriolipidosis-prone rats (ALR; Yamori 1984a) were selected for reactive hypercholesterolemia from among A sublines.

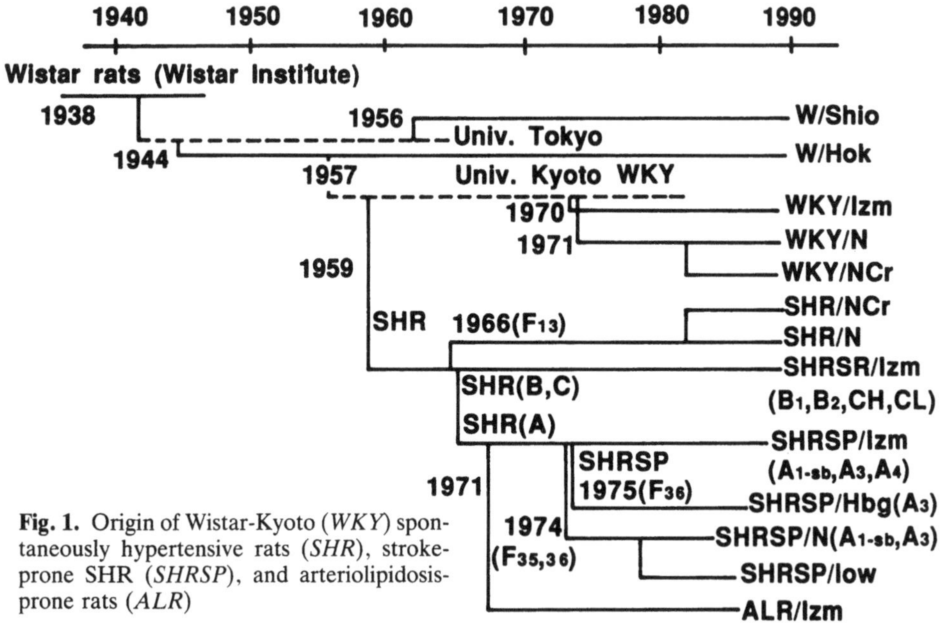

Fig. 1. Origin of Wistar-Kyoto (*WKY*) spontaneously hypertensive rats (*SHR*), stroke-prone SHR (*SHRSP*), and arteriolipidosis-prone rats (*ALR*)

Genetic Variability of SHR and WKY

Such genotypic variabilities were confirmed by Matsumoto et al.'s (1991) extensive studies on a total of 32 miscellaneous polymorphic enzymes and proteins. Among them, nonspecific esterases 3 and 4 in intestine and kidney, respectively, were found to be useful common markers of SHR strains in general, and pepsinogen 1 (pg-1) and 6-phosphogluconate dehydrogenase-1 (pgd-1) were useful markers for identifying SHRSP from SHRSR (Fig. 2).

Esterase-4 in both renal medulla and cortex, of which we first described the most clear allele difference between SHR and control WKY (Yamori and Okamoto 1970), is now known to be very closely linked with other esterases 2, 3, 7, 8, 9 and 10, and these esterases can be controlled by one or two genes (Matsumoto et al. 1991). Thirty two loci were so far examined; isozyme patterns of esterases 3 and 4 were identical in all SHR strains, both SHRSP and SHRSR available in the world and were different from those of WKY strains. The loci of pepsinogen-1 and phosphogluconate dehydrogenase in all SHRSP strains were clearly different from those in all SHRSR strains (Fig. 2).

Although linkage groups are known for some of these loci, chromosome numbers are not known for most, except for pgd-1 which is known to be located on chromosome No. 5. Allele differences in SHR and SHRSR were noted at aldehyde-dehydrogenase-2 and c and alkaline phosphatase-1, while SHRSP showed two allele differences at aldehyde dehydrogenase-c and serum alkaline phosphatase (Matsumoto et al. 1991). Interestingly, both SHRSR and SHRSP shared allele differences at similar loci as WKY strains.

LOCI

Strains	ACON1	ACP2	AHD2	AHDC	AKP1	ALP	AMY1	CS	ES1	ES2	ES3	ES4	ES6	ES7	ES8
SHR/Jim	-	-	b	b	-	a	-	-	a	a	b	a	a	b	b
SHR/Kyo	b	a	c	-	a	a	a	b	a	a	b	a	a	b	b
SHR/NCrj	b	a	b	a	a	a	-	b	a	a	-	a	a	b	b
SHRSRCL/Izm	b	a	c	b	a	a	a	b	a	a	b	a	a	b	b
SHRSRB1/Izm	b	a	b	b	a	a	a	b	a	a	b	a	a	b	b
SHRSRB2/Izm	b	a	c	b	b	a	a	b	a	a	b	a	a	b	b
SHRSRCH/Izm	b	a	c	b	a	a	a	b	a	a	b	a	a	b	b
SHRSP/Jim	-	-	b	a	-	b	-	-	a	a	b	a	a	b	b
SHRSP/Hkm	-	-	b	b	-	b	-	-	a	a	b	a	-	b	b
SHRSP/Kur	b	a	b	a	b	-	a	b	a	a	b	a	a	b	b
SHRSP/N	b	a	b	a	b	a	a	b	a	a	b	a	a	b	b
SHRSPA1/Izm	b	a	b	a	b	b	a	b	a	a	b	a	a	b	b
SHRSPA3/Izm	b	a	b	a	b	b	a	b	a	a	b	a	a	b	b
SHRSPA3H/Izm	b	a	b	a	b	-	a	b	a	a	b	a	a	b	b
SHRSPA4/Izm	b	a	b	a	b	b	a	b	a	a	b	a	a	b	b
WKY/Kur1	b	b	c	a	a	-	a	b	a	a	a	b	a	b	b
WKY/Kur2	b	b	c	a	a	-	a	b	b	a	d	b	a	b	b
WKY/NCrj	b	a	b	a	b	b	-	a	a	d	-	b	a	b	a
WKY/Slc	b	a	b	-	b	-	-	-	a	d	-	-	-	-	-
WKY/Izm	b	a	c	a	a	b	a	b	a	d	d	b	a	b	a
WKYO/Ham	b	a	b	a	b	b	a	a	a	d	a	b	a	b	a
WKYO/Kyo	b	a	b	a	b	a	a	b	a	d	d	b	a	b	a

Strains	ES9	ES10	ESSI	FH	GC	GDC1	GST1	HAO1	HBB	LAP1	MUP1	PEP3	PG1	PGD	SVP1
SHR/Jim	a	a	b	b	-	-	b	-	-	-	-	-	-	b	a
SHR/Kyo	a	a	b	b	a	a	-	a	a	a	a	a	-	b	a
SHR/NCrj	a	a	b	b	a	a	b	-	a	a	-	-	-	b	-
SHRSRCL/Izm	a	a	b	b	a	a	b	a	a	a	a	a	b	b	a
SHRSRB1/Izm	a	a	b	b	a	a	b	a	a	a	a	a	b	b	a
SHRSRB2/Izm	a	a	b	b	a	a	b	a	a	a	a	a	b	b	a
SHRSRCH/Izm	a	a	b	b	a	a	b	a	a	a	a	a	b	b	a
SHRSP/Jim	a	a	b	b	-	-	b	-	-	-	-	-	-	c	a
SHRSP/Hkm	a	a	-	b	-	-	-	-	-	-	-	-	-	c	a
SHRSP/Kur	a	a	-	b	a	a	-	a	a	a	-	a	a	-	-
SHRSP/N	a	a	b	b	a	a	-	a	a	a	a	a	a	c	a
SHRSPA1/Izm	a	a	b	b	a	a	b	a	a	a	-	a	a	c	a
SHRSPA3/Izm	a	a	b	b	a	a	b	a	a	a	a	a	a	c	a
SHRSPA3H/Izm	a	a	b	b	a	a	b	a	a	a	a	a	a	c	a
SHRSPA4/Izm	a	a	b	b	a	a	b	a	a	a	-	a	a	c	a
WKY/Kur1	a	a	-	b	a	a	-	a	a	a	-	b	b	-	a
WKY/Kur2	a	a	-	b	a	a	-	a	b	a	-	b	b	-	a
WKY/NCrj	c	b	b	b	a	a	a	a	a	a	-	-	-	c	-
WKY/Slc	-	b	-	b	a	-	-	a	-	-	-	-	-	-	-
WKY/Izm	c	b	b	b	a	a	a	a	a	a	a	b	b	c	a
WKYO/Ham	c	b	b	b	a	a	b	a	a	a	a	b	b	c	a
WKYO/Kyo	c	b	b	b	a	a	b	a	a	a	a	b	b	c	a

Fig. 2. Allele distribution among SHR, stroke-prone SHR (*SHRSP*), stroke-resistant SHR (*SHRSR*), and Wistar-Kyoto (*WKY*). Enzymes assayed: *Aconl*, aconitase-1; *ACP2*, acid phosphatase-2; *AHD2*, aldehyde dehydrogenase-2; *AHDC*, aldehyde dehydrogenase-c; *AKP1*, kidney alkaline phosphatase-1; *ALP*, serum alkaline phosphatase; *AMY1*, amylase-1; *CS*, red cell catalase; *ES1–ES10*, esterase-1–Esterase-10; *ESSI*, sex-influenced esterase; *FH*, fumarate hydratase; *GC*, group specific component; *GDC1*, glycerophosphate dehydrogenase; *GST1*, glutathione S-transferase-1; *HAO1*, hydroxyacid oxidase-1; *HBB*, hemoglobin beta chain; *LAP1*, leucine amino peptidase-1; *MUP1*, major (or male) urinary protein-1; *PEP3*, peptidase-3; *PG1*, pepsinogen-1; *PGD*, phosphogluconate dehydrogenase; *SVP1*, seminal vesicle protein-1. Isozyme pattern specific for SHRSP and SHRSR and strain difference between SHRSP and SHRSR. (Partly from Matsumoto et al. 1991): □

To compare the degree of strain variabilities, numbers of different alleles between the strains were counted: the greater the numbers, the greater the differences between the strains. The genetic variabilities within SHR or SHRSR strains are observed at 1 to 3 loci; the NIH strain obtained through Charles River appears identical with other SHR strains in Europe but shows differences at 1 to 3 loci from all SHR in Japan. The allele distribution among SHRSP strains was different at only two loci. By contrast, WKY strains show much genetic variability; for example, the NIH strain (WKY/N) (Department of Health and Human Services 1981) has allele differences at 10 loci from one WKY in Japan (WKY/Kur2). WKY strains show many differences from SHR strains in general, being different maximally at 13 and 11 loci from SHRSR and SHRSP, respectively. This is because the esterase gene cluster closely linked with each other are different between SHR and WKY. Therefore, substantial different loci between SHR and WKY strains are regarded as 3 to 6.

The genetic traits of these rats were further studied by using RT-1 class 1 polymorphism (Matsumoto et al. 1991). All SHRSR and SHRSRP strains as well as WKY in Japan and original Wistar rats available now in Japan have RT-1^K haplotype, without exception. However, the NIH strain of WKY and the Charles River strain have RT-1^L haplotype, different from WKY in Japan. Since the product of this gene cluster determines self or non-self, the WKY strain with closer haplotype with SHR should be genetically closer to SHR and be a better control for the genetic analysis on hypertensinogenic genes.

DNA Polymorphisms in Search of Hypertensinogenic Genes

Genetic variabilities among WKY and SHR have been further confirmed by our recent DNA fingerprint analysis (Kurtz et al. 1989; Nabika et al., 1991; Samani et al. 1989). Among WKY from three sources, the NIH strain and the Charles River strain derived from the NIH strain during the inbreeding process showed very similar patterns which were obviously different from the patterns of WKY in Izumo (WKY/Izm). DNA fingerprints obtained from the liver, spleen and tail of WKY/Izm and WKY/N were identical within the strain but different from each other.

Similarly SHRSR from both NIH and Charles River showed very similar patterns because of their common origin but were substantially different from four strains of SHRSR/Izm. These four strains showed differences among themselves, since they were separated from each other at the F_{13} generation and thereafter at our laboratory.

In contrast to SHRSR strains, the SHRSP of NIH (Department of Health and Human Services 1981) and Izumo strains (SHRSP/N, SHRSP/Izm) had similar fingerprinting patterns. However, two distinct, different bands were observed among three strains of SHRSP/Izm which were separated from each other at the F_{21} generation. Since our preliminary study suggested these bands were two different alleles of the same locus, this difference might be generated by mutation before the present F_{80} generation (Nabika et al., 1991).

We further compared DNA fingerprints between our WKY and seven SHRSR and SHRSP strains after genomic DNA was digested by Hae III and was hybridized with 'myo,' composed of tandem repeats of synthesized oligomers whose sequences were identical to the repetitive oligonucleotide sequences in human myoglobin gene (Nabika et al., submitted). WKY/Izm had a fingerprinting pattern close to SHR/Izm strains – one band at about 3.7 Kbp was commonly noted in all seven SHR strains – whereas WKY/Izm did not have this band. Such genetic loci showing restriction fragment length polymorphisms (RFLPs) between WKY and all SHR strains, but not among all SHR strains, are expected to have linkage with hypertension loci with greater possibility. Similarly, DNA fingerprints, after genetic DNA was digested by Hinf 1 and hybridized with 'ins' of repetitive oligonucleotide sequences in human insulin gene, demonstrated SHRSP- and SHRSR-specific fingerprinting bands (Nabika et al., 1991). It is clear that segregation studies of the F_2 or back cross generation are needed to reach any conclusions as to whether these genetic differences are really involved in genetic causes of hypertension or stroke.

F_2 Analysis of Genetic Traits

F_1 and F_2 segregate generations were repeatedly produced by cross-breeding between normotensive strains and SHRSR to investigate the mode of inheritance and the number of genes involved in hypertension or stroke (Tanase et al. 1970; Yamori et al. 1972). As shown in one example of recent cross-breeding experiments between SHRSP and WKY/Izm (Fig. 3), age-related blood

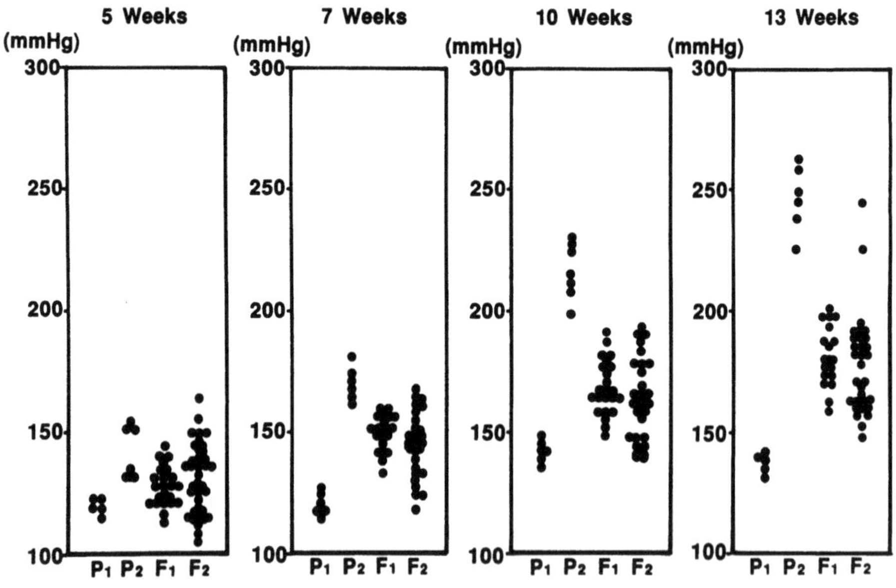

Fig. 3. Distribution of blood pressure in F_1 and F_2 generations obtained by cross-breeding between WKY (P_1) and SHRSP (P_2) at 5–13 weeks of age

pressure rise with some variances in F_1 and F_2 generations was observed within the wide blood pressure range between two parent strains.

The means of blood pressure in F_1 and F_2 generations were close to, but always somewhat lower than, the mid parent value, indicating some dominant influence of normotensive gene(s).

Since the variance in F_1 hybrids (VF_1) with homogeneous genotype is due to environmental influence (VE), and that in F_2 generations (VF_2) is due to both genetic and environmental factors (VG + VE), the degree of genetic determination (DGD), the ratio of genetic variance to total variance, can be calculated from these genetic and environmental variances by the following formula (Tanase et al. 1970);

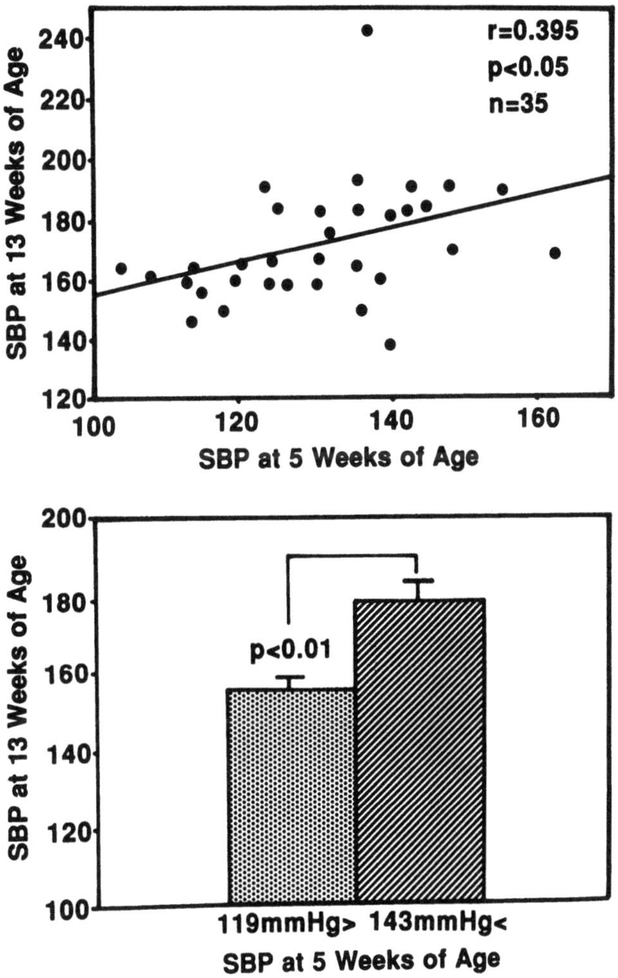

Fig. 4. Tracking phenomenon of blood pressure in F_2 generation obtained from cross-breeding between WKY and SHRSP. Correlation between systolic blood pressures (SBP) at 5 weeks of age and those at 13 weeks of age (*above*), and SBP levels at 13 weeks of age of two groups with highest and lowest blood pressure quinteil (*below*)

$$DGD = VG/(VG + VE) = (VF_2 - VF_1)/VF_2$$

DGD calculated from blood pressure variances in F_1 and F_2 generations from 5 through 13 weeks of age were high, around 50 to 70%, and confirmed our previous results (Tanase et al. 1970). The minimum number of major genes involved in hypertension calculated from blood pressure distribution in F_2 was around three, as speculated previously (Tanase et al. 1970).

We further analyzed tracking in blood pressure in F_2 generation. Significant correlations in blood pressures were observed between blood pressures at 5 and 7 weeks of age, between those at 7 and 10 weeks of age and also between those at 10 and 13 weeks of age. Therefore, blood pressures in F_2 obtained by the first measurement at 5 weeks of age were significantly related to blood pressures at 13 weeks of age (Fig. 4). Individuals in the highest and lowest quinteil of blood pressure distribution at 5 weeks of age showed significant blood pressure differences at 13 weeks of age. As already observed in the F_2 segregate generation, an early rise in blood pressure can be a good predictor of hypertension in later life.

Several characteristics of SHRSP, such as salt sensitivity, salt preference and behavioral alertness, did not cosegregate with blood pressure in the F_2 generation. The SHR strain-specific allele difference in esterase-4 showed only weak correlation with blood pressure in F_2 (Fig. 5); the F_2 with the SHR esterase pattern showed significantly higher blood pressure than F_2 with the WKY pattern, but the difference in average blood pressure was small (Yamori et al.

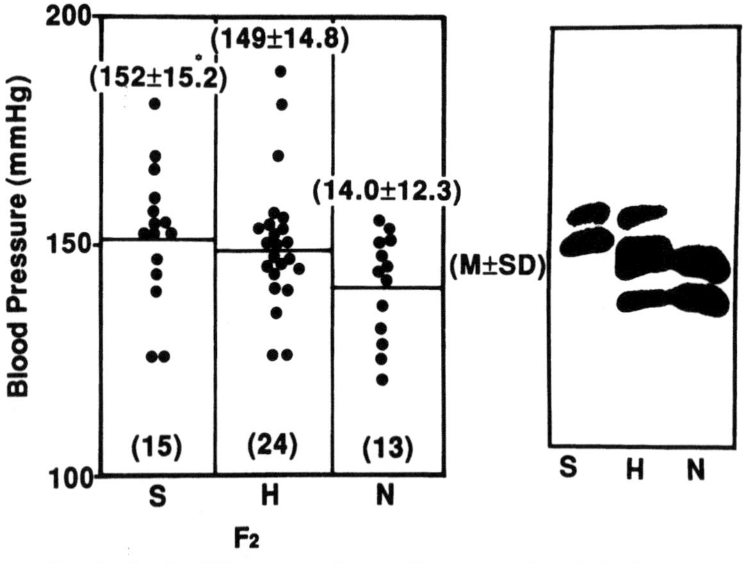

Fig. 5. Relationship between blood pressure and renal esterase-4 isozyme patterns in F_2 generation obtained by cross-breeding between SHR and WKY. *S, H,* and *N: SHR,* hybrid, and normal (WKY) patterns, respectively. (Data partly from Yamori et al. 1972)

1972). Similarly, cosegregation with blood pressure in F_2 is now being analyzed for the aforementioned several RFLPs noted in SHR strains. If any close linkage is detected, such loci could be used for studying genetic pathogenesis or for predicting hypertension.

Pathogenesis and Predictors of Hypertension, Atherosclerosis and Stroke

Pathogenesias and Predictors of Hypertension

Although detection of hypertensinogenic genes should be the most reliable way to clarify pathogenesis and establish predictive methods, pathophysiological studies on SHR up to the present indicate that neurogenic, vascular and membrane factors are involved in the pathogenesis, and early rise in blood pressure, exaggerated response to stress, cardiovascular hypertrophy and some membrane transport abnormalities can be used as predictors for hypertension (Yamori 1983, 1989a).

Among these predictors, tracking of blood pressure confirmed in F_2 can also be used as one of the most practical predictors for hypertension in man. A cohort study on school children (Shimane Heart Study) showed that both girls and boys whose systolic blood pressures were in the highest or the lowest quinteil at the age of 12 belonged more frequently to the highest or the lowest quinteil at the age of 15 (Nishio et al. 1987).

Neurogenic factors in SHR can be detected at an early stage, as early as four weeks after birth even before the development of hypertension (Yamori et al. 1985). A cold exposure test clearly induces a marked rise in urinary epinephrine excretion in SHR and SHRSP. Since this rise is completely blocked after adrenal denervation, this exaggerated response to cold exposure is elicited by the central mechanisms. The same findings were noted in young normotensive volunteers at the average age of 24 (Yamori 1989a). Cold exposure provoked a greater rise in urinary epinephrine excretion in the young volunteers with a positive family history of hypertension either on low salt or high salt diets (Fig. 6).

Cardiac hypertrophy in hypertension has long been regarded as the most common secondary response to hypertension. However, studies on SHR demonstrated that heart weight as well as vector cardiogram were hypertrophic even before blood pressure was elevated significantly (Yamori et al. 1979).

Since vascular hypertrophy was concomitantly observed with cardiac hypertrophy, the growth rate of vascular smooth muscle cells (VSMC) was examined under tissue culture conditions. VSMC obtained from the aorta of SHR and SHRSP showed a significantly rapid growth without any influence of blood pressure (Yamori 1983; Yamori et al. 1981a, 1988). Therefore, we can conclude that VSMC of hypertensive rats are genetically different in growth rate.

Although the early clinical detection of vascular hypertrophy is not yet possible, early development of cardiac hypertrophy can be detected echocardiographically in young children in primary and secondary schools the size of the heart in children with higher blood pressures over 90, 66 and 50 percentiles was

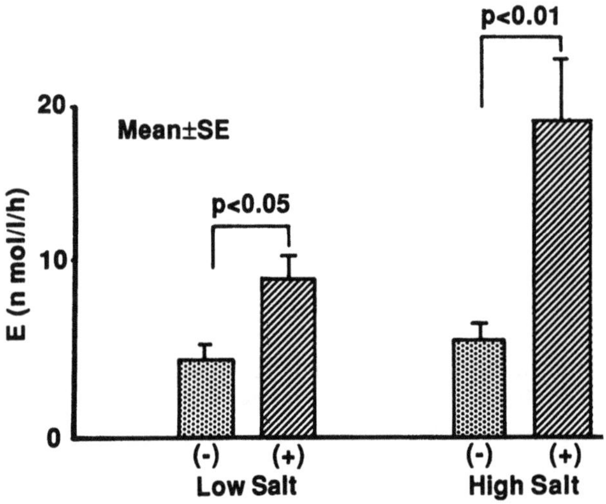

Fig. 6. Cold exposure-induced urinary epinephrine (*E*) excretion in normotensive young males with, (+), or without (−), family history of hypertension during high (25 g/day) and low (6 g/day) salt feeding. (Data from Yamori 1989a)

always significantly greater than those with lower blood pressures below 10, 33 and 50 percentiles, respectively (Yamori 1989a; Yamori et al. 1979; Nishio et al. 1979). Therefore, early development of cardiac hypertrophy seems to be a predictor for the development of hypertension in later life.

As for salt sensitivity, defined as blood pressure rise induced by salt loading, we first noted that SHRSP were more salt sensitive than SHRSR and WKY strains (Yamori et al. 1981b). Our studies on cultured VSMC demonstrated that salt-sensitive SHRSP showed increased permeability of Na and K ions through the membrane of VSMC (Yamori 1983; Yamori et al. 1982) and more accelerated Na^+-H^+ exchange as determined by the speed of intracellular pH changes in response to intracellular acidification (Fig. 7; Kobayashi et al. 1990). Such biomembrane or ionic transport abnormalities may contribute to the accelerated growth rate of VSMC as one of the hypertension mechanisms and may also be used as cellular predictors of hypertension.

Pathogenesis and Predictors of Atherosclerosis

Such accelerated growth of VSMC may also be involved in the genetic pathogenesis of atherosclerosis. A greater reactive hypercholesterolemia was noted especially in ALR, hypertensive models for atherosclerosis which were developed as one of SHRSP strain (SHRSP $A_{1\text{-sb}}$; Yamori 1984) from the same ancestor SHR (A) subline. Increased LDL incorporation into cultured VSMC appears to be an important cellular mechanism of atherosclerosis in ALR and SHR. Increased sudanophilic droplets containing cholesterol ester accumulated more rapidly into cells from SHRSP than those from normotensive WKY, when LDL was added to culture medium (Yamori et al. 1987). These results suggest that cellular abnormalities related not only to hypertension but also to atherosclerosis can be detected for predicting the development of atherosclerosis.

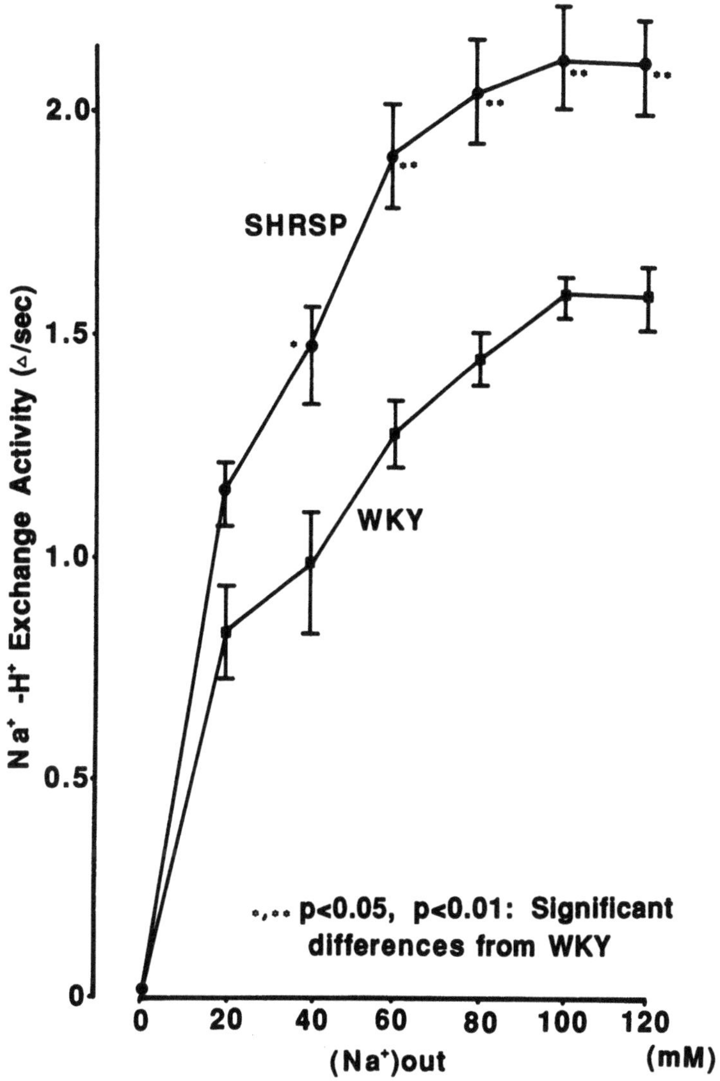

Fig. 7. Na^+-H^+ exchange activity of culture vascular smooth muscle cells from the aortas of SHRSP and WKY at different Na concentrations in the medium. (Data from Kobayashi et al. 1990)

Pathogenesis and Predictors of Stroke

As for genetic pathogenesis and predictors of stroke, vulnerability to hypoxia of cultured VSMC (Yamori 1989a; Yamori et al. 1987), as well as osmotic fragility of erythrocytes previously noted in SHRSP and human with a predisposition to stroke (Yamori et al. 1980), may be an important cellular basis for the development of cerebrovascular lesions. Regional cerebral blood flow reduction was a pathophysiologically important cause of cerebrovascular lesions both for hemorrhagic and ischemic stroke in SHRSP (Yamori 1989a; Yamori and Horie 1977). Predilection sites of stroke in the brain of SHRSP as well as in man were fed by perforating arteries which branched off rather recurrently from the main blood stream. Thus, hypertension rheologically reduces blood flow and the entry of erythrocyte-carrying oxygen into these perforating arteries. Moreover, these arteries have no vasa vasorum which supply arterial wall with blood from the adventitia.

In addition, nutrition should be supplied from the lumen to the vascular wall through tightly connected endothelial cells with a blood-brain barrier function. Therefore, VSMC in the media initially show focal degeneration and necrosis at the outer layer (Yamori 1989a; Tagami et al. 1987), the part furthest from blood-derived oxygen and nutritional diffusion; then degeneration and necrosis spread over the whole layer of the media. This morphological observation indicates that the basic process in cerebrovascular damage in hypertension is due to local nutritional disturbance.

A similar process has been observed in autopsied cases of cerebral hemorrhage, infarction or cerebrovascular dementia from which brain specimens were taken within five hours after death and fixed in methanol-Carnoy solution and embedded in paraffin. Monoclonal antibody (HHF35; Gown et al. 1986) against VSMC actin applied to these specimens successfully demonstrated that intact VSMC with actin remained only at the inner layer, and the outer layer was totally replaced with collagen (Yamori 1989a), whereas monoclonal antibody agianst human macrophages (HAM56) demonstrated the infiltration of positively stained macrophages into the outer layer of the media and adventitia from the Virchow-Robin space at the initial stage and the subendothelial invasion at the advanced stage.

Mechanisms of stroke are summarized in Figure 8 according to our new concepts, based on our experimental findings in SHRSP and our immunohistological findings on autopsied stroke cases.

Hypertension induces reduction of cerebral blood flow which aggravates local nutritional disturbance in VSMC of small intracerebral arteries and accelerates degeneration or necrosis at the outer layer of the media. When macrophages initially invade into necrotized vascular wall from outside the arterial wall as scavenger cells, their fibroblast-stimulating factor accelerates collagen synthesis and the process of arteriosclerosis. When macrophages invade through the endothelial cell layer and destroy the blood-brain barrier at the advanced stage, fibrin deposits in arterial walls and results in cerebral hemorrhage when the walls are ruptured, or in cerebral infarction with thrombosis.

Therefore, the basic process of vascular lesions in stroke is regarded as VSMC degeneration and necrosis of intracerebral arteries caused by local nutritional disturbance. This new concept about the pathogenesis of stroke is strongly supported by our previous observations on effective nutritional prevention of stroke in SHRSP.

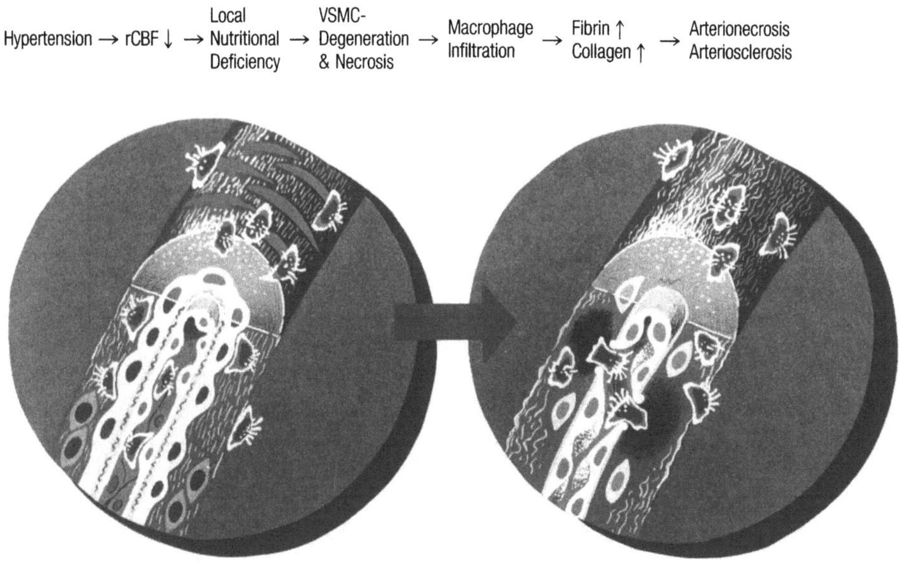

Fig. 8. Schematic drawing of mechanisms of cerebrovascular diseases in hypertension

Prevention of Cardiovascular Diseases for Longevity

The preventive effect of a protein-rich diet on stroke was first demonstrated in our experiment in SHRSP (Yamori 1983, 1989a; Yamori et al. 1976). When they were fed a normal or low protein diet with excess salt, the SHRSP developed severe hypertension quickly and all died of stroke within a short period. Without excess salt 80 % of them died of stroke. In contrast, when fed a high protein diet with excess salt, the incidence of stroke was markedly reduced, and in the group fed a high fish protein diet without excess salt intake, the development of severe hypertension was attenuated and no stroke was observed. Thus, we can conclude that a high protein diet attenuates the development of severe hypertension and counteracts the adverse effects of salt (Yamori 1981).

We further confirmed the beneficial effects of high protein diets in man and also the mechanism and beneficial amino acid components of protein (Yamori 1989a; Yamori et al. 1984).

Among various beneficial dietary factors, such as potassium, Ca, Mg, dietary fibers and some fatty acids previously proven to be effective in SHRSP (Yamori 1989a) Figure 9 shows remarkable dietary preventive effect of protein, Ca and Mg supplementation in SHRSP (Yamori et al. 1989). When they were fed on soy protein-rich diets, or Ca-rich diets or Mg-rich diets, the survival rate of SHRSP was significantly improved in comparison with SHRSP given normal diets with 1 % salt in drinking water. These results indicate that these individual dietary components can reduce the adverse effects of salt, partly by accelerating urinary Na excretion as in the case of K-rich diets (Yamori et al. 1984, 1989). Moreover, the combination of protein and Ca, and protein and Ca plus Mg, successfully prevented stroke and markedly improved the survival rate.

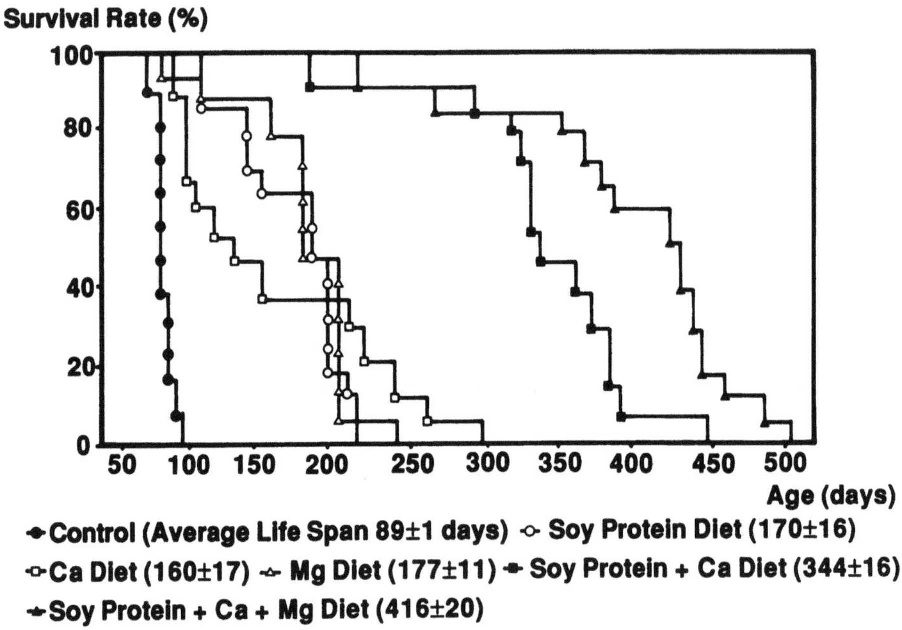

Fig. 9. Effects of dietary soy protein, Ca, and Mg supplementations on survival rates of SHRSP given 1% salt water for drinking

We can now conclude that dietary components are important and effective for preventing stroke in our animal models, and the applicability of this conclusion to the prevention of CVD in man has been proven by our international cooperative studies on cardiovascular diseases and alimentary comparison, the so-called WHO CARDIAC Study (Yamori 1989a; Cardiac Study Group 1989; Yamori 1989b; Yamori et al. 1990), presently covering 48 populations in 21 countries.

Conclusion

Hypertension and related CVD develop under gene-environmental interactions (Fig. 10). Our recent studies suggest that genetic hypertension mechanisms as

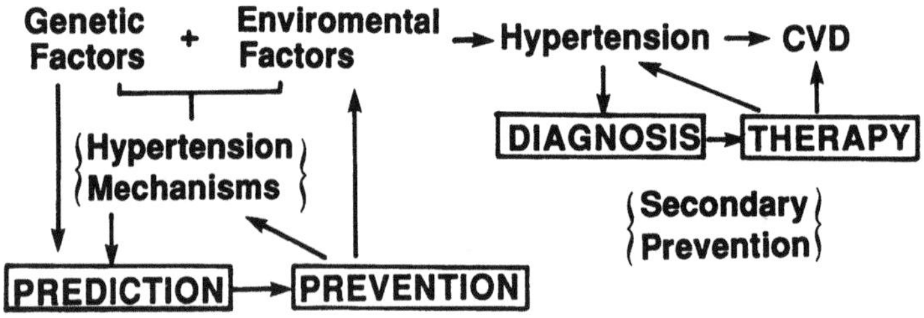

Fig. 10. Predictive and preventive medicine for hypertension-related cardiovascular diseases (*CVD*)

well as hypertensinogenic genes may be detectable early, before the development of hypertension. On the other hand, hypertensive diseases have been experimentally proven to be preventable by controlling environmental factors, mainly diets, even in genetically strongly predisposed animal models.

Although the main activities in medicine up to now have been "diagnosis" and "therapy" of diseases already developed, medical care should be started early, with the detection of causative genes in each healthy individual, and the main effort of medicine should be directed toward the complete prevention of diseases before gene-carriers develop diseases.

This is the best message obtained from our studies on animal models.

References

CARDIAC Study Group (1989) Purpose and design of the CARDIAC Study: Hypertension and biological dietary markers in urine and blood-A progress from the CARDIAC Study Group. In: Yamori Y, Strasser T (eds) New horizons in preventing cardiovascular diseases. Elsevier, Amsterdam, pp 111–126

Department of Health and Human Services (1981) NIH Rodents: 1980 Catalogue. (NIH publication no 81–606)

Gown AM, Tsukuda T, Ross R (1986) Human atherosclerosis: immunocytochemical analysis of the cellular composition of human atherosclerotic lesions. Am J Pathol 125: 191–207

Kobayashi A, Nara Y, Nishio T, Mori C, Yamori Y (1990) Increased Na^+-H^+ exchange activity in cultured vascular smooth muscle cells from spontaneously hypertensive rats. J Hypertens 8: 153–157

Kurtz TW, Montano M, Chan L, Kabra P (1989) Molecular evidence of genetic heterogeneity in Wistar-Kyoto rats: implications for research with the spontaneously hypertensive rat. Hypertension 13: 188–192

Matsumoto K, Yamada T, Natori T, Ikeda K, Yamada J, Yamori Y (1991) Genetic variability in SHR (SHRSR, SHRSP) and WKY strains. Clin Exp Hypertens A13 in press

Nabika T, Nara Y, Ikeda K, Endo J, Yamori Y (1991) Genetic variability of the spontaneously hypertensive rats. Hypertension 18: 12–16

Nishio T, Mori C, Saito M, Soeda T, Abe K, Nakao Y (1979) Left ventricular hypertrophy in early hypertensive children: its importance as a risk factor for hypertension. Shimane J Med Sci 3: 71–79

Nishio T, Mori C, Haneda N, Watanabe K, Kishida K, Saito M, Kajino Y (1987) Quantification of blood pressure tracking of children by tracking index: the Shimane Heart study. Jpn Circ J 51: 1404–1408

Okamoto K, Aoki K (1963) Development of a strain of spontaneously hypertensive rats. Jpn Circ J 27: 282–293

Samani NJ, Swales JD, Jeffreys AJ, Morton DB, Naftilan AJ, Lindpaintner K, Ganten D, Brammar WJ (1989) DNA fingerprinting of spontaneously hypertensive and Wistar-Kyoto rats: implications for hypertension research. J Hypertens 7: 809–816

Tagami M, Nara Y, Kubota A, Sunaga T, Maezawa H, Fujino H, Yamori Y (1987) Ultrastructural characteristics of occluded perforating arteries in stroke-prone spontaneously hypertensive rats. Stroke 18: 733–740

Tanase H, Suzuki Y, Ooshima A, Yamori Y, Okamoto K (1970) Genetic analysis of blood pressure in spontaneously hypertensive rats. Jpn Circ J 34: 1197–1212

Yamori Y (1981) Environmental influences on the development of hypertensive vascular diseases in SHR and related models, and their relation to human diseases. In: Worcel M, Bonvalet JP, Langer SZ, Menard J, Sassard J (eds) New trends in arterial hypertension. Elsevier, Amsterdam, pp 305–320

Yamori Y (1983) Physiopathology of the various strains of spontaneously hypertensive rats. In: Genest J, Kuchel O, Hamet P et al. (eds) Hypertension. McGraw-Hill, Montreal, pp 556–581

Yamori Y (1984a) Development of the spontaneously hypertensive rats (SHR) and of various spontaneous rat models, and their implications. In: de Jong W (ed) Experimental and genetic models of hypertension. Elsevier, Amsterdam, pp 224–239 (Handbook of hypertension, vol 4)

Yamori Y (1984b) The stroke-prone spontaneously hypertensive rats: contribution to risk factor analysis and prevention of hypertensive diseases. In: de Jong W (ed) Experimental and genetic models of hypertension. Elsevier, Amsterdam, pp 240–255 (Handbook of hypertension, vol 4)

Yamori Y (1989a) Prediction and preventive pathology of cardiovascular diseases. Acta Pathol Jpn 39: 683–705

Yamori Y (on behalf of CARDIAC Study Group) (1989b) Preliminary report of CARDIAC Study: Cross-sectional multicenter study on dietary factors of cardiovascular diseases. Clin Exp Hypertens [A] 11: 957–972

Yamori Y, Horie R (1977) Developmental course of hypertension and regional cerebral blood flow in stroke-prone spontaneously hypertensive rats. Stroke 8: 456–461

Yamori Y, Okamoto K (1970) Zymoigram analysis of various organs from spontaneously hypertensive rats. Lab Invest 22: 206–211

Yamori Y, Ooshima A, Okamoto K (1972) Genetic factors involved in spontaneously hypertension in rats: an analyis of F_2 segregate generation. Jpn Circ J 36: 561–568

Yamori Y, Nagaoka A, Okamoto K (1974) Importance of genetic factors in hypertensive cerebovascular lesions: an evidence obtained by successive selective breeding of stroke-prone and -resistant SHR. Jpn Circ J 38: 1095–1100

Yamori Y, Horie R, Akiguchi I, Ohtaka M, Nara Y, Fukase M (1976) New models of SHR for studies on stroke and atherogenesis. Clin Exp Pharmacol Physiol 3 [Suppl]: 199–203

Yamori Y, Mori C, Nishio T, Ooshima A, Horie R, Ohtaka N, Soeda T, Saito M, Abe K, Nara Y, Nakao Y, Kihara M (1979) Cardiac hypertrophy in early hypertension. Am J Cardiol 44: 964–969

Yamori Y, Nara Y, Horie R, Ooshima A (1980) Abnormal membrane characteristics of erythrocytes in rat models and men with predisposition to stroke. Clin Exp Hypertens 2: 1009–1021

Yamori Y, Igawa T, Kanbe T, Kihara M, Nara Y, Horie R (1981a) Mechanisms of structural vascular changes in genetic hypertension: Analyses on cultured vascular smooth muscle cells from spontaneously hypertensive rats. Clin Sci 61 [Suppl 7]: 121s–123s

Yamori Y, Nara Y, Kihara M, Horie R, Ooshima A (1981b) Sodium and other dietary factors in experimental and human hypertension – the Japanese experience. In: Laragh JH, Bühler FR, Seldin DW (eds) Frontiers in hypertension research. Springer, Berlin Heidelberg New York, pp 46–48

Yamori Y, Nara Y, Kanbe T, Imafuku H, Mori K, Kihara M, Horie R (1982) Diversity of membrane abnormalities in spontaneously hypertension. Clin Sci 63 [Suppl 8]: 27s–29s

Yamori Y, Horie R, Nara Y, Kihara M, Fujiwara K, Nabika T, Mano M, Ikeda K (1984) Nutritional causation and prevention of cardiovascular diseases-experimental evidence in animal models and man. In: Lovenberg W, Yamori Y (eds) Nutritional prevention of cardiovascular disease. Academic, Orlando, pp 37–51

Yamori Y, Ikeda K, Kulakowski EC, McCarty R, Lovenberg W (1985) Enhanced sympathetic-adrenal medullary response to cold exposure in spontaneously hypertensive rats. J Hypertens 3: 63–66

Yamori Y, Horie R, Nara Y, Tagami M, Kihara M, Mano M, Ishino H (1987) Pathogenesis and dietary prevention of cerebrovascular diseases in animal models and epidemiological evidence for the applicability in man. In: Yamori Y, Lefant C (eds) Prevention of cerebrovascular diseases: An approach to active long life. Elsevier, Amsterdam, pp 163–177

Yamori Y, Igawa T, Kanbe T, Nara Y, Tagami M (1988) Enhanced growth rate of cultured vascular smooth muscle cells from spontaneously hypertensive rats. Heart Vessels 4: 94–99

Yamori Y, Nara Y, Ikeda K, Tsuchikura S, Eguchi T, Mano M, Horie R, Sugahara T (1989) Recent advances in experimental studies on dietary prevention of cardiovascular diseases. In: Yamori Y, Strasser T (eds) New horizons in preventing cardiovascular diseases. Elsevier, Amsterdam, pp 1–11

Yamori Y, Nara Y, Mizushima S, Mano M, Sawamura M, Kihara M, Horie R, Hatano S (1990) International cooperative study on the relationship between dietary factors and blood pressure: a report from the cardiovascular diseases and alimentary comparison study. J Cardiovasc Pharmacol 16 [Suppl 8]: s43–s47

An Overview of the Genetics of Coronary Heart Disease and Its Risk Factors

K. Berg

Summary

Aggregation in families of cases of premature coronary heart disease (CHD), as well as a high level of heritability of several risk factors or protective factors with respect to CHD, show that genes contribute significantly to susceptibility or resistance to atherosclerotic disease. Current attempts to identify single genes contributing to the susceptible state focus on the candidate gene approach. This approach was already being used in the early 1970s when a definite association was detected between a genetically determined high level of Lp(a) lipoprotein and CHD. The availability of DNA technology has greatly widened the scope of the candidate gene approach. Associations to lipid levels have been found with normal DNA variants at the low density lipoprotein (LDL) receptor locus and at the apolipoprotein B (apoB) locus.

It is possible that both absolute risk factor levels and amount of risk factor variability play a role in atherogenesis. Gene-environment interaction must be important in the etiology of CHD, life style or dietary factors preferentially causing disease in those with a genetic predisposition. Recent evidence indicates that gene-gene interaction also influences risk factor variation. It is anticipated that ongoing research will greatly improve the possibilities for predicting CHD risk. Efficient prediction of disease risk should become an important tool in a system of family-oriented preventative medicine. The strong evidence that nutritional and life style factors are important in the etiology of coronary heart disease (CHD) and the limited attention paid until recently to genetic factors have led to the misconception that CHD is almost exclusively caused by environmental factors. Curiously, the careful documentation in the 1930s that the triad of tendon xanthomas, early CHD and hypercholesterolemia behaves as an autosomal dominant trait in many families (Müller 1939) had little influence for many years on the thinking concerning CHD etiology. Apparently, genetic factors may be ignored once the importance of environmental factors has been established for a disease.

K. Berg, V. Bulyzhenkov, Y. Christen, P. Corvol (Eds.)
Genetic Approaches to Coronary Heart Disease
and Hypertension

Clustering of Coronary Heart Disease Cases in Families

Familial clustering of coronary heart disease case has been analyzed by several workers. Particularly important studies were conducted on Finnish (Rissanen 1979a, b; Rissanen and Nikkilä 1977, 1979; Nikkilä and Aro 1973) and US (Nora et al. 1980) populations. Evidence emerged in both countries for a strong effect of genes where CHD had occurred at a relatively young age. In Finland, practically all cases of premature CHD had strong genetic components to their etiology. In most cases, the interpretation was that the etiology was polygenic. However, when the index case was younger than 45 years, the distribution of affected and healthy persons in many families pointed to a monogenic etiology, even in the absence of classical monogenic hyperlipidemia. The risk to relatives of the youngest index cases was the same in the high-incidence area of East Finland as in areas where the incidence of CHD is much lower. Thus, genetic factors in these cases obliterated the effects of atherogenic diet and life style in East Finland.

In the Denver study (Nora et al. 1980) of CHD occurring before age 55, a heritability of 0.56 was observed even after patients with monogenic hyperlipidemias had been excluded. Having a first-degree relative with CHD prior to age 55 was a risk factor in its own right, as was having a second-degree relative with CHD. The risk caused by having a first-degree relative with CHD prior to age 55 was higher than the sum of the risks caused by cigarette smoking and moderately increased cholesterol level.

Several other studies are in agreement with the conclusion that genetic factors are of considerable importance for CHD occurring prior to age 55–60 in males or 60–65 in females. The author (Berg 1979, 1983, 1987a–c) and others have previously reviewed the literature extensively.

Genetics of Risk Factor Levels

More than 20 years ago it became evident that the population distribution of cholesterol levels is influenced by genes (Mayo et al. 1969, 1971). Cholesterol levels are influenced by genes in several animal species, and in recent years genetic influence on several risk factors or protective factors with respect to CHD has been uncovered (for review, see Berg 1979, 1983, 1987a–c, 1990a). Heritability estimates from our own studies are shown in Table 1, which summarizes the results of two twin studies (Berg 1990b). In the first twin study (study 1) monozygotic (MZ) as well as dizygotic (DZ) twins were examined and classical twin analysis was conducted. In the second study, only MZ twins were examined and the within-pair correlation was taken as an estimate of heritability. The heritability figures for apolipoproteins proved strikingly similar between the two studies. Blood pressure, body mass index (Berg 1990b) and obesity (Stunkard et al. 1986) also exhibited impressive heritability (data not shown). These observations illustrate the need to identify single genes involved in determining levels of CHD risk factors or protective factors.

Table 1. Heritability (h^2) of serum lipid and apolipoprotein (apo) levels in a series of Norwegian twins consisting of 98 monozygotic (MZ) and 100 same-sexed dizygotic (DZ) pairs (study 1), and in a series of 156 MZ twin pairs (study 2). In the latter study, the intraclass correlation coefficient in MZ twins was used as a heritability estimate. (From Berg 1987a)

Parameter	Heritability (h^2) estimate	
	Study 1	Study 2
Cholesterol	0.34	0.68
Fasting triglycerides	0.40	0.46
apoB level	0.66	0.64
apoA-I level	0.53	0.55
apoA-II level	0.69	0.68

Markers of Genetic Coronary Heart Disease Risk

Early attempts to study the genetics of CHD and risk factors made use of formal genetic analyses, the methods of quantitative genetics, and the use of random genetic markers (in the hope that some of the marker genes would be close to functionally important genes). Current attempts focus on the "candidate gene" approach. With respect to atherosclerosis and CHD there are several categories of candidates, such as genes whose protein products are:
- involved in lipoprotein structure, apolipoprotein metabolism or lipid metabolism
- involved in thrombogenesis, thrombolysis or fibrinolysis
- involved in regulation of blood flow in coronary arteries
- involved in regulation of blood pressure
- involved in reverse cholesterol transport
- present in atherosclerotic lesions
- involved in the regulation of growth of atherosclerotic lesions
- involved in the early development of coronary arteries

This approach was applied in the 1970s, at the lipoprotein level, when a clear association between genetically determined Lp(a) lipoprotein and CHD was established (Berg et al. 1974), as was association between lipid levels and genetic variants in the Ag allotypic system of low density lipoprotein (LDL; Berg et al. 1976). The genetically determined isoforms of apolipoprotein E (apoE) were also shown to be associated with cholesterol levels, with the apoE4 isoform significantly raising total and LDL cholesterol levels (Utermann et al. 1979).

Recombinant DNA technology has added a new dimension to the study of candidate genes with respect to CHD risk. Apolipoprotein genes have been cloned and genetic variation has been detected at their loci. Studies on the relationship between normal DNA variation at apolipoprotein loci and risk factor levels have uncovered association between cholesterol and triglyceride levels on one hand, and a restriction fragment length polymorphism (RFLP) at the apolipoprotein B (apoB) locus detectable with the restriction enzyme XbaI on the other (Law et al. 1986; Berg 1986). This RFLP is caused by a silent mutation in the third base of the codon corresponding to amino acid 2488 in the mature protein. Accordingly, this association should reflect genetic variation in a functionally important domain of the apoB gene which is in linkage disequilibrium with the RFLP.

There is strong association between Ag allotypic LDL variants and DNA variants at the apoB locus (Berg 1986), and linkage analyses (Berg et al. 1986) have proved that the Ag variants reside in apoB rather than in non-protein parts of LDL. Significantly, the association between Ag allotypes and the XbaI RFLP is consistent with the associations between Ag variants and lipids, as well as with the associations between DNA variants and lipids (Berg et al. 1986; Berg 1986).

Studies of DNA variation at the apoB locus have also suggested direct association with CHD (Hegele et al. 1986) as well as with lipids and obesity (Rajput-Williams et al. 1988). There are many reports of associations between risk factor levels and normal variation at apolipoprotein loci, but several of the associations have been difficult to confirm.

Studies of candidate genes other than apolipoprotein genes have also lead to interesting results. Thus, Pedersen and Berg (1988) found association between an RFLP at the LDL receptor (LDLR) locus and total as well as LDL cholesterol. This finding was in excellent agreement with reports several years ago that genes affect LDLR function in the normal population and suggestive evidence that such effects influence the population variation in total cholesterol (Maartmann-Moe et al. 1981). To rule out false positive associations, we reexamined the association between cholesterol levels and DNA variants at the LDLR locus and confirmed it in a second series (Pedersen and Berg 1990). This observation has since been independently confirmed (D. A. Coviello, personal communication). Thus, normal genetic variants at the LDLR locus contribute to the population variation in total and LDL cholesterol. The RFLP studied resides in an intron and could by itself not have a functionally important role. Its association with total and LDL cholesterol levels must be caused by linkage disequilibrium with functionally important domains of the LDLR gene.

DNA variation at still another nonlipoprotein candidate locus may be related to level of CHD risk factors or protective factors. "Reverse cholesterol transport" is the least studied and poorest understood part of lipid metabolism. This area became much more accessible to study when Drayna et al. (1987) cloned the gene for cholesteryl ester transfer protein (CETP) and later detected RFLPs at its locus (Drayna and Lawn 1987). In a population study, evidence emerged for association between levels of high density lipoprotein (HDL) cholesterol as well as apoA-I and an RFLP at the CETP locus (Kondo et al. 1989).

The study of "candidate genes" is still at an early stage. Nevertheless, this approach will probably contribute to the identification of most genes involved in CHD susceptibility or resistance.

The Lp(a) Lipoprotein

The Lp(a) lipoprotein (Berg 1963, 1968, 1979, 1983) is a class of serum lipoprotein particles that received only limited attention until the gene coding for the polypeptide chain carrying the Lp(a) antigen(s) (the Lp(a) polypeptide chain or the Lp(a) apolipoprotein) was cloned by workers in San Francisco and Chicago (McLean et al. 1987), and extensive homology with plasminogen was detected.

The genetic nature of the Lp(a) lipoprotein variation had already been uncovered in the early studies (Berg 1963, 1968) and has been confirmed in numerous studies since. Analysis of Lp(a) lipoprotein as a quantitative character has strongly supported single-locus control of high Lp(a) lipoprotein levels (Schultz et al. 1974; Sing et al. 1974; Morton et al. 1985), and very close genetic linkage with plasminogen (Weitkamp et al. 1988; Drayna et al. 1988; Berg 1989a, 1990c) has proved beyond any doubt that Lp(a) lipoprotein level is controlled by one single locus, the LPA locus (for review, see Berg 1990c, d).

Association between premature CHD and Lp(a) lipoprotein was detected in 1974 (Berg et al. 1974) and has since been repeatedly confirmed (Brown and Goldstein 1987). Rhoads et al. (1986) found a population attributable risk of 28% for myocardial infarction in men below age 60 who had an Lp(a) lipoprotein level in the top quartile of the population distribution. The corresponding figure for males 60–69 years of age was 13%.

Durrington et al. (1988) found that practically all familial aggregation of early CHD cases in the absence of monogenic hyperlipidemias was caused by a high Lp(a) lipoprotein level. The Lp(a) lipoprotein level is independent of variation in LDL or other lipid or apolipoprotein parameters and is, in healthy people, a constant trait. Accordingly, a high Lp(a) lipoprotein level is a genetic risk factor that can be detected early in life (for example in early adult life) for the purpose of disease prevention. These and other results make it important to include Lp(a) lipoprotein analysis in any population screening for CHD risk. Furthermore, such analyses should be mandatory in the diagnostic work-up of patients with early CHD.

The mechanism underlying the association between genetically determined, high level of Lp(a) lipoprotein and CHD is not fully known. However, Walton (1972) detected apparently intact Lp(a) lipoprotein in atherosclerotic lesions, and Dahlén et al. (1978) found excessive flocculation of Lp(a) lipoprotein particles under conditions mimicking those in arterial walls. Rath et al. (1989) confirmed that Lp(a) lipoprotein is deposited in atherosclerotic lesions and demonstrated direct association between serum concentration and amount deposited in arterial walls.

The recent detection of extensive homology between the polypeptide chain carrying the Lp(a) antigen(s) and plasminogen suggests additional mechanisms to explain the Lp(a) lipoprotein-CHD association. Harpel et al. (1989) observed affinity between Lp(a) lipoprotein and protease-modified fibrinogen or fibrin; Hajjar et al. (1989) demonstrated Lp(a) lipoprotein modulation of endothelial cell surface fibrinolysis. Miles et al. (1989) detected Lp(a) lipoprotein competition for plasminogen receptors by molecular mimicry, and we observed a (relatively weak) association between Lp(a) lipoprotein and fibrinogen levels. Thus, in vitro evidence suggests that the association between CHD and a high Lp(a) lipoprotein level may (at least in part) be explained by interference of Lp(a) lipoprotein with thrombolytic/fibrinolytic processes.

The existence of isoforms of the polypeptide chain carrying the Lp(a) antigen(s) (Fless et al. 1984; Utermann et al. 1987) and the danger that antiserum may cross-react with plasminogen make quantitative immunoelectrophoresis the method of choice for Lp(a) lipoprotein measurements. The use of commercially offered kits that are based on other techniques is questionable. Furthermore, monoclonal antibodies should at present not be used, since they may not react equally well with all Lp(a) lipoprotein isoforms. The somewhat laborious technique that presently is the method of choice for Lp(a) lipoprotein measurement will doubtless limit the use of Lp(a) lipoprotein testing for clinical purposes. However, determination of Lp(a) lipoprotein level would not have to be repeated numerous times since it is a constant trait. The importance of high-quality antisera cannot be over-emphasized. There are at present few sources for high-quality antisera to Lp(a) lipoprotein but they do exist. Only reagents from these soures can be recommended.

Gene-Environment Interactions and the Variability Gene Concept

In several animal species, there are high responders and low responders with respect to blood lipid reaction to fat intake. These traits are strain specific and almost certainly genetically determined. There are suggestions from clinical studies in man that the amount of variability in serum cholesterol may be of importance for the risk to develop atherosclerotic disease (Groover et al. 1960). Katan and coworkers (Beynen and Katan 1985; Katan and Beynen 1987; Katan et al. 1986) have shown that high responders and low responders to lipid intake exist in man, and that the traits persist at least over many years. It is plausible that these traits are inherited, as they are in several animal species. Apparently, the same change in lipid intake (Table 2) can produce widely different reactions with respect to blood lipid levels, depending on host-specific factors that are very likely to reflect genetic differences.

Several years ago, we realized that there was a need to develop genetic approaches to risk factor analysis that were more dynamic than the search for

Table 2. Cholesterol response to a given dietary cholesterol load according to susceptibility to diet. (From Katan et al. 1986)

Group	Response (mmol/l)
The 16% who are least susceptible	⩽0.29
The 16% who are most susceptible	⩾0.87
Mean	0.58

associations between absolute risk factor levels and genetic markers. We developed and have applied a method aimed at detecting genes ("variability genes") contributing to the framework within which dietary or life style factors can cause risk factor variation. Presumably, such genes could be different from genes ("level genes") contributing to absolute risk factor levels as judged from classical association tests. Our method is based on the fact that MZ twins have identical genes so that any difference in a given quantitative parameter between the two members of a pair must be caused by environmental factors. A permissive "variability gene" would lead to a higher within-pair difference in MZ pairs possessing the gene than in MZ pairs lacking it, whereas a restrictive "variability gene" would have the opposite effect (Berg 1981, 1984, 1988, 1990a). Using random genetic markers, we demonstrated several years ago that the "variability gene" concept is plausible (Magnus et al. 1981; Berg 1981, 1984) and we have recently confirmed variability gene effects of genes at or near the locus for the Kidd blood group on lipid levels (Berg 1988). Studying DNA variation at many candidate loci, we have observed several suggestions of "variability gene" effects (Berg 1984, 1988, 1990a; Berg et al. 1989).

The potential clinical importance of the "variability gene" concept resides in the likelihood that a person's total genetic risk of CHD depends on his or her *combination* of level genes and variability genes, as indicated in Table 3. One of the most intriguing possibilities is that people with different combinations of risk factors or protective factors may need different preventative efforts from those recommended for the general population.

Gene-Gene Interaction

The postulated interaction between "level genes" and "variability genes", as well as between these categories of genes and nutritional or life style factors, are not the only interactions that deserve further study with respect to hyperlipidemia or CHD risk. It has been known for several years (Utermann et al. 1979) that the genetically determined apoE isoforms influence lipid levels to a considerable extent. We have previously reported (Berg 1987a) a population sample from Norway where the difference in mean serum cholesterol between

Table 3. Genetic risk for coronary heart disease (CHD) resulting from the combination of genes affecting level and variability, respectively, of CHD risk factors. (Adapted from Berg 1987a)

Risk factor level specified by level genes	Total genetic risk if variability genes are:	
	Permissive	Restrictive
High	High, but reducible	Very high
Average	Average, but changeable	Average
Low	Low, but changeable	Very low

Table 4. Sex- and age-adjusted total cholesterol (TC) and LDL cholesterol (LDLC) in 239 unrelated healthy Norwegians according to genotype in the apolipoprotein E (apoE) polymorphism

apo E genotype	*N*	Mean TC (mmol/l)	Mean LDLC (mmol/l)
E2E2	4	4.99	3.24
E2E3	33	5.56	3.58
E2E4	2	6.58	4.43
E3E3	149	5.83	3.94
E3E4	49	6.43	4.53
E4E4	2	6.36	4.54

apoE4 and apoE2 homozygotes was 1 mmol/l. A second series of healthy Norwegians was recently examined, and the strong effect of the apoE4 isoform on total serum and LDL cholesterol was confirmed (Table 4).

The series where the effect of normal genes at the LDLR locus on cholesterol level was first detected (see above) had also been examined with respect to apoE isoforms. The cholesterol-increasing effect of the apoE4 isoform was only found (Pedersen and Berg 1989) in people who lacked a normal gene at the LDLR locus, expressed as presence of a PvuII restriction site (Table 5). This effect was confirmed (Pedersen and Berg 1990) in a second series of unrelated individuals.

Table 5. Gene-Gene interaction between apolipoprotein E (apoE) alleles and normal alleles at the low density lipoprotein receptor (LDLR) locus expressed as a *Pvu*II restriction site polymorphism, in determining age- and sex-adjusted total cholesterol. (From Pedersen and Berg 1989)

	Mean total cholesterol (mmol/l)		
apoE4 allele	In people with LDLR genotype		In total series
	A2A2	A1A1[a] or A1A2[b]	
Present	7.06[c] (n = 31)	5.87 (n = 15)	6.67 (n = 46)
Absent	6.08[c] (n = 69)	5.84 (n = 41)	5.99 (n = 110)

[a] A1 allele, presence of *Pvu*II site.
[b] A2 allele, absence of *Pvu*II site.
[c] $t = 3.90, p < 0.001$

We conclude that the first gene-gene interaction in determining the level of a CHD risk factor in the population at large has been uncovered and we predict that other gene-gene interactions are likely to be found.

We cannot at present offer a plausible explanation for the observed interaction. A potentially important consequence of this interaction could be that, if one attempts to predict future cholesterol level from apoE typing, one would have a sizable fraction of wrong results unless genes at the LDLR locus were examined at the same time.

Family-Oriented Preventative Medicine

Existing and forthcoming knowledge will doubtless lead to greatly improved possibilities to identify people with a genetic CHD susceptibility. The purpose of identifying at-risk individuals would be to improve disease prevention by particular efforts in those at the highest risk. It is a reasonable expectation that preventative efforts will be more effective the earlier they are started and that a strong motivation will help the individual in adhering to preventative regimens. These considerations and the high percentage of people with a genetic CHD risk would, by themselves, be strong reasons to offer risk screening to young adults. If screening is voluntary, and the results of tests are fully confidential, such screening could become an important tool in preventative cardiology aimed at the individual.

It would be particularly important to analyze close relatives (particularly offspring) of people with premature myocardial infarction, with respect to detectable CHD predisposition. The aim of such analyses would be to institute preventative measures in close relatives who are found to be at risk, but who are still healthy at the time of examination. One would aim at manipulating all risk factors that can at present be changed.

For the above purposes it may be useful to develop a family-oriented preventative medicine (Berg 1989b) where the nuclear family is the target as well as the executive instrument. The risk in the family should be a strong motivating factor, and disease prevention and health promotion should become an important, joint project for the whole family.

Acknowledgments. Work in the author's laboratory is supported by The Norwegian Council on Cardiovascular Diseases, the Norwegian Research Council for Science and the Humanities and Anders Jahres Foundation for the Promotion of Science, as well as by grant No. EDC-1 1 R01 HD26746-01 GT from the National Institute of Child Health and Human Development, National Institutes of Health, Bethesda, Maryland. Part(s) of this paper was written while the author was a Scholar-in-Residence, Fogarty International Center for Advanced Studies in the Biomedical Science, National Institutes of Health, Bethesda, Maryland, USA.

References

Berg K (1963) A new serum type system in man – the Lp system. Acta Pathol Microbiol Scand 59: 369–382

Berg K (1968) The Lp system. Ser Haematol 1: 111–136

Berg K (1979) Inherited lipoprotein variation and atherosclerotic disease. In: Scanu AM, Wissler RW, Getz GS (eds) The biochemistry of atherosclerosis. Dekker, New York, pp 419–490

Berg K (1981) Twin research in coronary heart disease. In: Gedda L, Parisi P, Nance WE (eds) Twin research 3: part C, epidemiological and clinical studies. Liss, New York, pp 117–130

Berg K (1983) Genetics of coronary heart disease. In: Steinberg AG, Bearn AG, Motulsky AG, Childs B (eds) Progress in medical genetics. Saunders, Philadelphia, pp 35–90

Berg K (1984) Twin studies of coronary heart disease and its risk factors. Acta Genet Med Gemellol 33: 349–361
Berg K (1986) DNA polymorphism at the apolipoprotein B locus is associated with lipoprotein level. Clin Genet 30: 515–520
Berg K (1987a) Genetic risk factors for atherosclerotic disease. In: Vogel F, Sperling K (eds) Human genetics. Proceedings of the 7th international congress. Springer, Berlin Heidelberg New York, pp 326–335
Berg K (1987b) Genetics of coronary heart disease and its risk factors. In: Bock G, Collins GM (eds) Molecular approaches to human polygenic disease. Wiley, Chichester, pp 14–33 (Ciba Foundation Symposium 130)
Berg K (1987c) Genetics of atherosclerosis. In: Olsson AG (ed) Atherosclerosis. Biology and clinical science. Churchill-Livingstone, Edinburgh, pp 323–337
Berg K (1988) Variability gene effect on cholesterol at the Kidd blood group locus. Clin Genet 33: 102–107
Berg K (1989a) Close linkage between Lp(a) lipoprotein and plasminogen. Cytogen Cell Genet 51 (1–4): 961
Berg K (1989b) Impact of medical genetics on research and practices in the area of cardiovascular disease. Clin Genet 36: 299–312
Berg K (1990a) Level genes and variability genes in the etiology of hyperlipidemia and atherosclerosis. In: Berg K, Retterstøl N, Refsum S (eds) From phenotype to gene in common disorders. Munksgaard, Copenhagen, pp 77–91
Berg K (1990b) Molecular genetics and nutrition. In: Simopoulos AP, Childs B (eds) Genetic variation and nutrition. Karger, Basel, pp 49–59 (World rev Nutr Diet vol. 63)
Berg K (1990c) Genetics of atherogenic Lp(a) lipoprotein. In: Berg K, Retterstøl N, Refsum S (eds) From phenotype to gene in common disorders. Munksgaard, Copenhagen, pp 138–162
Berg K (1990d) Lp(a) lipoprotein: an overview. In: Scanu A (ed) Lipoprotein (a). Academic, New York, pp 1–23
Berg K, Dahlén G, Frick MH (1974) Lp(a) lipoprotein and pre-β_1-lipoprotein in patients with coronary heart disease. Clin Genet 6: 230–235
Berg K, Hames C, Dahlén G, Frick MH, Krishan I (1976) Genetic variation in serum low density lipoprotein and lipid levels in man. Proc Natl Acad Sci USA 73: 937–940
Berg K, Powell LM, Wallis SC, Pease R, Knott TJ, Scott J (1986) Genetic linkage between the antigenic group (Ag) variation and the apolipoprotein B gene: Assignment of the Ag locus. Proc Natl Acad Sci USA 83: 7367–7370
Berg K, Kondo I, Drayna D, Lawn R (1989) "Variability gene" effect of cholesteryl ester transfer protein (CETP) genes. Clin Genet 35: 437–445
Beynen AC, Katan MB (1985) Reproducibility of the variations between humans in the response of serum cholesterol to cessation of egg consumption. Atherosclerosis 57: 19–31
Brown MS, Goldstein JL (1987) Teaching old dogmas new tricks. Nature 330: 113–114
Dahlén G, Ericson C, Berg K (1978) In vitro studies of the interaction of isolated Lp(a) lipoprotein and other serum lipoproteins with glycosaminoglycans. Clin Genet 14: 36–42
Drayna D, Lawn R (1987) Multiple RFLPs at the human cholesteryl ester transfer protein (CETP) locus. Nucleic Acids Res 15: 4698
Drayna D, Jarnagin AS, McLean J, Henzel W, Kohr W, Fielding CJ, Lawn R (1987) Cloning and sequencing of human cholesteryl ester transfer protein cDNA. Nature 327: 632–634
Drayna DT, Hegele RA, Hass PE, Emi M, Wu LL, Eaton DL, Lawn RM, Williams RR, White RL, Lalouel J-M (1988) Genetic linkage between lipoprotein(a) phenotype and a DNA polymorphism in the plasminogen gene. Genomics 3: 230–236
Durrington PN, Hunt L, Ishola M, Arrol S, Bhatnagar D (1988) Apolipoproteins (a), AI, and B and parental history in men with early onset ischaemic heart disease. Lancet i: 1070–1073
Fless GM, Rolih CA, Scanu AM (1984) Heterogeneity of human plasma lipoprotein(a). J Biol Chem 259: 11470–11478
Groover ME, Jernigan JA, Martin CD (1960) Variations in serum lipid concentration and clinical coronary disease. Am J Med Sci 53: 133–139

Hajjar KA, Gavish D, Breslow JL, Nachman RL (1989) Lipoprotein (a) modulation of endothelial cell surface fibrinolysis and its potential role in atherosclerosis. Nature 339: 303–305
Harpel PC, Gordon BR, Parker TS (1989) Plasmin catalyzes binding of lipoprotein (a) to immobilized fibrinogen and fibrin. Proc Natl Acad Sci USA 86: 3847–3851
Hegele RA, Huang LS, Herbert PN, Blum CB, Buring JE, Hennekens CH, Breslow JL (1986) Apolipoprotein B-gene DNA polymorphisms associated with myocardial infarction. N Engl J Med 315: 1509–1515
Katan MB, Beynen AC (1987) Characteristics of human hypo- and hyperresponders to dietary cholesterol. Am J Epidemiol 125: 387–399
Katan MB, Beynen AC, De Vries JHM, Nobels A (1986) Existence of consistent hypo- and hyperresponders to dietary cholesterol in man. Am J Epidemiol 123: 221–234
Kondo I, Berg K, Drayna D, Lawn R (1989) DNA polymorphism at the locus for human cholesteryl ester transfer protein (CETP) is associated with high density lipoprotein cholesterol and apolipoprotein levels. Clin Genet 35: 49–56
Law A, Powell LM, Brunt H, Knott TJ, Altman DG, Rajput J, Wallis SC, Pease RJ, Priestley LM, Scott J, Miller GJ, Miller NE (1986) Common DNA polymorphism within the coding sequence of the apolipoprotein B gene associated with altered lipid levels. Lancet i: 1301–1303
Maartmann-Moe K, Magnus P, Golden W, Berg K (1981) Genetics of the low density lipoprotein receptor: III. Evidence for multiple normal alleles at the low density lipoprotein receptor locus. Clin Genet 20: 113–129
Magnus P, Berg K, Børresen A-L, Nance WE (1981) Apparent influence of marker genotypes on variation in serum cholesterol in monozygotic twins. Clin Genet 19: 67–70
Mayo O, Fraser GR, Stamatoyannopoulos G (1969) Genetic influences on serum cholesterol in two Greek villages. Hum Hered 19: 86–99
Mayo O, Wiesenfeld SL, Stamatoyannopoulos G, Fraser GR (1971) Genetical influences on serum-cholesterol level. Lancet ii: 554–555
McLean JW, Tomlinson JE, Kuang W-J, Eaton DL, Chen EY, Fless GM, Scanu AM, Lawn RM (1987) cDNA sequence of human apolipoprotein (a) is homologous to plasminogen. Nature 330: 132–137
Miles LA, Fless GM, Levin EG, Scanu AM, Plow EF (1989) A potential basis for the thrombotic risks associated with lipoprotein (a). Nature 339: 301–303
Morton NE, Berg K, Dahlén G, Ferrell RE, Rhoads GG (1985) Genetics of the Lp lipoprotein in Japanese-Americans. Genet Epidemiol 2: 113–121
Müller C (1939) Angina pectoris in hereditary xanthomatosis. Arch Intern Med 64: 675–700
Nikkilä EA, Aro A (1973) Family study of serum lipids and lipoproteins in coronary heart disease. Lancet i: 954–959
Nora JJ, Lortscher RH, Spangler RD, Nora AH, Kimberling WJ (1980) Genetic-epidemiologic study of early-onset ischemic heart disease. Circulation 61: 503–508
Pedersen JC, Berg K (1988) Normal DNA polymorphism at the low density lipoprotein receptor (LDLR) locus associated with serum cholesterol level. Clin Genet 34: 306–312
Pedersen J, Berg K (1989) Interaction between low density lipoprotein receptor (LDLR) and apolipoprotein E (apoE) alleles contributes to normal variation in lipid level. Clin Genet 35: 331–337
Pedersen J, Berg K (1990) Normal genetic polymorphism at the LDL receptor locus and lipid levels. In: Berg K, Retterstøl N, Refsum S (eds) From phenotype to gene in common disorders. Munksgaard, Copenhagen, pp 107–114
Rajput-Williams J, Wallis SC, Yarnell J, Bell GI, Knott TJ, Sweetnam P, Cox N, Miller NE, Scott J (1988) Variation of apolipoprotein-B gene is associated with obesity, high blood cholesterol levels, and increased risk of coronary heart disease. Lancet ii: 1442–1446
Rath M, Niendorf A, Reblin T, Dietel M, Krebber H-J, Beisiegel U (1989) Detection and quantification of lipoprotein(a) in the arterial wall of 107 coronary bypass patients. Arteriosclerosis 9: 579–592
Rhoads GG, Dahlén G, Berg K, Morton NE, Dannenberg AL (1986) Lp(a) lipoprotein as a risk factor for myocardial infarction. JAMA 256: 2540–2544

Rissanen AM (1979a) Familial occurrence of coronary heart disease: effect of age at diagnosis. Am J Cardiol 44: 60–66
Rissanen AM (1979b) Familial aggregation of coronary heart disease in a high incidence area (North Karelia, Finland). Br Heart J 42: 294–303
Rissanen AM, Nikkilä EA (1977) Coronary artery disease and its risk factors in families of young men with angina pectoris and in controls. Br Heart J 39: 875–883
Rissanen AM, Nikkilä EA (1979) Aggregation of coronary risk factors in families of men with fatal and non-fatal coronary heart disease. Br Heart J 42: 373–380
Schultz JS, Shreffler CD, Sing CF (1974) The genetics of the Lp antigen. I. Its quantitation and distribution in a sample population. Ann Hum Genet 38: 39–46
Sing CF, Schultz JS, Shreffler DC (1974) The genetics of the Lp antigen. II. A family study and proposed models of genetic control. Ann Hum Genet 38: 47–56
Stunkard AJ, Sørensen TIA, Hanis C, Teasdale TW, Chakraborty R, Schull WJ, Schulsinger F (1986) An adoption study of human obesity. N Engl J Med 314: 193–198
Utermann G, Pruin N, Steinmetz A (1979) Polymorphism of apolipoprotein E. III. Effect of single polymorphic gene locus on plasma lipid levels in man. Clin Genet 15: 63–72
Utermann G, Menzel HJ, Kraft HG, Duba HC, Kemmier HG, Seitz C (1987) Lp(a) glycoprotein phenotypes. J Clin Invest 80: 458–465
Walton KW (1972) Distribution of lipoproteins in arteries and tissues as determined by immunohistological and radioactive tracer methods. In: Peeters H (ed) Protides of the biological fluids, 19th Colloquium 1971. Pergamon, Oxford, pp 225–226
Weitkamp LR, Guttormsen SA, Schultz JS (1988) Linkage between the loci for the Lp(a) lipoprotein (LP) and plasminogen (PLG). Hum Genet 79: 80–82

Genetic Risk Factors for Coronary Heart Disease in the Japanese Population

H. Hamaguchi, S. Kikuchi, H. Yanagi, T. Matsunaga,
Y. Yamanouchi, A. Nakagawa, Y. Watanabe, T. Tanaka,
Y. Shimakura, N. Imoto, T. Arinami,
and *K. Yamakawa-Kobayashi*

Summary

Our recent data on genes associated with premature coronary heart disease (CHD) or possibly involved in CHD in the Japanese population are described. Frequency distributions, mean values and median values of plasma lipoprotein (a) (Lp(a)) concentrations in a Japanese population are similar to those in most Caucasians. Preliminary data suggest that apolipoprotein (a) is also highly polymorphic in the Japanese population. Furthermore, in inactive plasminogen, phenotype M5, is polymorphic in the Japanese. Data obtained with polymerase chain reaction (PCR) and allele-specific oligonucleotide probes indicate that plasminogen M5 has a codon 601 missense mutation (GCT → ACT, Ala → Thr) and is identical with plasminogen Tochigi. As to apolipoprotein (apo) E polymorphisms, data obtained with PCR and allele-specific oligonucleotide probes suggest that many of apo E-2 have the codon 158 missense mutation (CGC → TGC, Arg → Cys) and that most, if not all, of apo E-4 have the codon 112 missense mutation (TGC → CGC, Cys → Arg) in the Japanese. The frequencies of alleles for apo E-2 and E-4, however, is lower in Japanese than in most Caucasians. As regarding familial hypercholesterolemia (FH), whose heterozygote frequency is estimated to be 1 in 500 persons in the Japanese, data obtained with family studies of the low density lipoprotein (LDL) receptor gene suggest that most, if not all, of relatively severe hereditary hypercholesterolemia associated with Achilles tendon xanthomas is caused by a defect of the LDL receptor gene, and that the origin of the mutant LDL receptor gene generally differs among different pedigrees in the Japanese population. On the other hand, data obtained by population and family studies suggest that hereditary moderate hypercholesterolemia due to abnormal gene(s) distinct from the abnormal LDL receptor gene may be more common than classic FH. As to the apolipoprotein A-I (apoA-I) gene, a mutant apoA-I gene with a codon 84 nonsense mutation which caused apoA-I deficiency, a marked reduction of HDL cholesterol and premature CHD is described. This finding suggests that a mutant apoA-I gene can act as a major gene for decreased levels of apoA-I and HDL cholesterol. Finally, application of genetic data for the prevention of premature CHD are discussed.

K. Berg, V. Bulyzhenkov, Y. Christen, P. Corvol (Eds.)
Genetic Approaches to Coronary Heart Disease
and Hypertension

Introduction

Genes associated with premature coronary heart disease (CHD) have been detected and characterized at the loci for apolipoprotein(a) (apo(a)), apolipoprotein E (apoE), low density lipoprotein (LDL) receptor, and apolipoprotein B (apoB; Utermann 1989; Mahley and Rall 1989; Goldstein and Brown 1989; Soria et al. 1989). Among them, genes at the loci for apo(a) and apoE are polymorphic. On the other hand, deleterious genes so far detected at the loci for LDL receptor and apoB are not polymorphic but act as major genes. Besides these loci, and inactive plasminogen, phenotype M5, has been reported with the gene frequency of about 0.02 in the Japanese population (Yamaguchi et al. 1989). Furthermore, we have recently revealed the presence of a nonsense mutation at the locus for apolipoprotein A-I (apoA-I) in a Japanese female patient with apoA-I deficiency and premature CHD (Matsunaga et al., submitted for publication). Our data obtained as a result of a population and family study, also suggest that hereditary moderate hypercholesterolemia due to abnormal gene(s) distinct from the abnormal LDL receptor gene is more common than classic familial hypercholesterolemia (FH).

We have been analyzing genes associated with premature CHD or possibly involved in premature CHD in the Japanese population, because it is important to elucidate genotypes associated with premature CHD for the prevention of this disease. In this communication, we report the results of our genetic studies of apo(a), plasminogen, apoE, autosomal dominant hypercholesterolemia due to the abnormal LDL receptor genes, hereditary moderate hypercholesterolemia caused by abnormal gene(s) other than the abnormal LDL receptor gene, and a deficiency of apoA-I and high density lipoprotein (HDL) cholesterol due to a mutant apoA-I gene.

Lp(a), Apolipoprotein(a) and Plasminogen M5 (Tochigi, 601 Ala → Thr)

Numerous epidemiological studies have found a positive association of plasma lipoprotein(a) (Lp(a)) concentrations with premature CHD (Utermann 1989). Lp(a) is made up of an LDL-like structure in which apolipoprotein B-100 is covalently linked to apo(a), which exhibits a striking similarity to plasminogen (Scanu and Fless 1990). The apo(a) locus is the major locus that determines plasma Lp(a) concentrations (Utermann 1989).

Figure 1 shows frequency distributions of plasma Lp(a) concentrations in 203 Japanese male adults with a mean age of 52.6 ± 8.5 and 158 children (80 boys and 78 girls) with a mean age of 10.6 ± 0.8. The subjects were unrelated and apparntly healthy. Lp(a) levels were measured using enzyme immunoassay for determination of human Lp(a) (Biopool AB, Umea, Sweden). As in Caucasians (Uterman 1989), serum Lp(a) levels vary more than 100-fold between Japanese individuals and represent a continuous quantitative trait with a skewed distribution in the Japanese population. In our data presented in Fig. 1, mean concentrations were 16.4 mg/dl in male adults and 15.8 mg/dl in children.

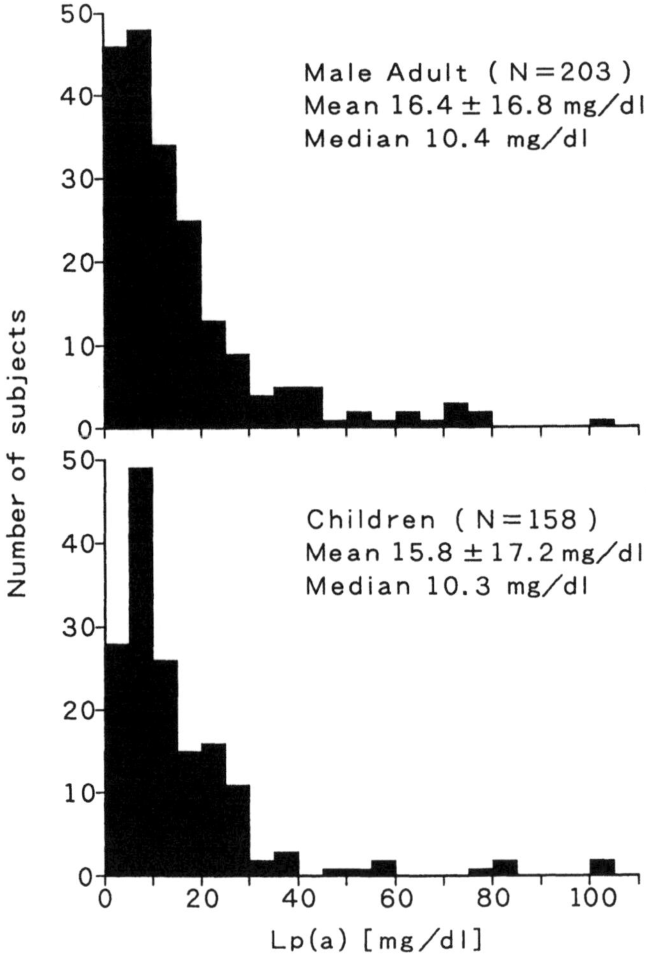

Fig. 1. Frequency distributions of plasma Lp(a) concentrations in 203 Japanese male adults and 158 children

Median concentrations were 10.4 mg/dl in male adults and 10.3 mg/dl in children. These values are similar to mean and median concentrations in most Caucasians (Utermann 1989). Percentage Lp(a) levels above 50 mg/dl were 5.9 % in adult males and 5.1 % in children.

Recently we prepared a monoclonal anti-apo(a) antibody which does not react with plasminogen. Our preliminary data obtained with immunoblotting using the monoclonal anti-apo(a) antibody suggest that more than ten apo(a) isoforms are present in the Japanese population. As in Caucasians (Utermann 1989; Gaubatz et al. 1990), apo(a) are very likely highly polymorphic in the Japanese.

As to the mechanism(s) for a cardiovascular pathogen of high plasma Lp(a) levels, it is apparent that Lp(a) may have both atherogenic and thrombogenic roles, the former through its LDL-like characteristic and the latter through the plasminogen-like properties of apo(a) (Scanu and Fless 1990). In this report, it is interesting that an inactive plasminogen (phenotype M5) is polymorphic in the Japanese population (Yamaguchi et al. 1989). The plasminogen activity of the heterozygote for the phenotype M5 is about half the normal activity and the frequency of the heterozygote is about 4%. Figure 2 shows plasminogen phenotypes A and A/M5 detectable with isoelectric focusing and immunoblotting. Plasminogen phenotype A is the most common and functionally normal.

We posited that plasminogen M5 is identical with plasminogen Tochigi (601 Ala → Thr) reported by Miyata et al. (1982) and have been testing this hypothesis using polymerase chain reaction (PCR) and dot blot hybridization with allele-specific oligonucleotide probes. Figure 3 shows the sequence of plasminogen phenotypes A and M5 in codon 601 and near the region estimated based upon the reported data (Miyara et al. 1982; Petersen et al. 1990). So far, we have examined eight alleles of plasminogen M5 derived from unrelated Japanese, and have observed that all eight alleles for the phenotype M5 have

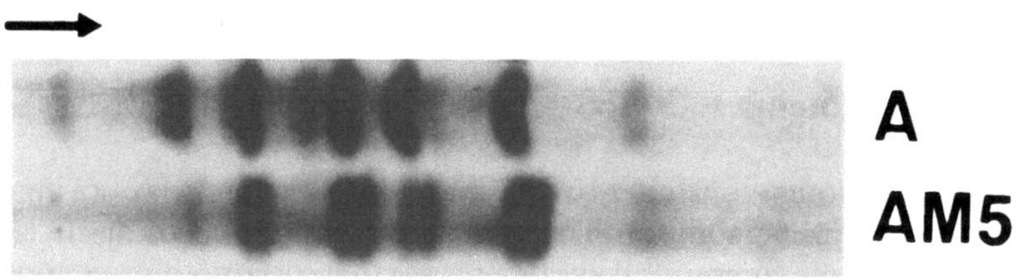

Fig. 2. Isoelectric focusing patterns of plasminogen phenotypes A and A/M5

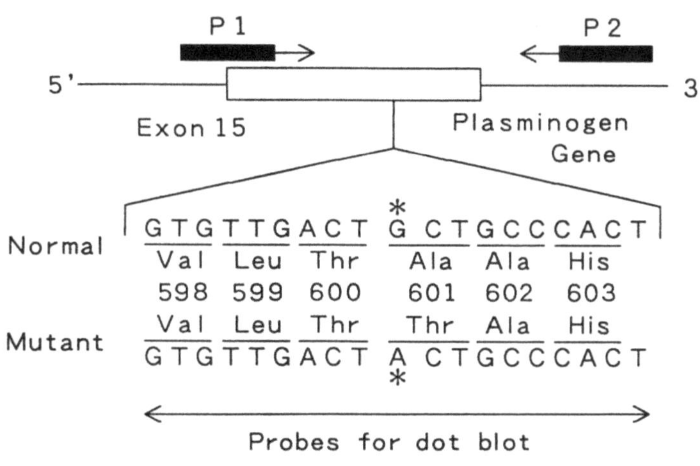

Fig. 3. Partial DNA and amino acid sequences of normal (phenotype A) and mutant (M5, Tochigi) plasminogen. The G to A substitution in codon 601 of exon 15 is shown by the *asterisk*. The positions of primers (*P1, P2*) for PCR and the sequences for dot blot hybridization with allele-specific oligonucleotide probes are also shown

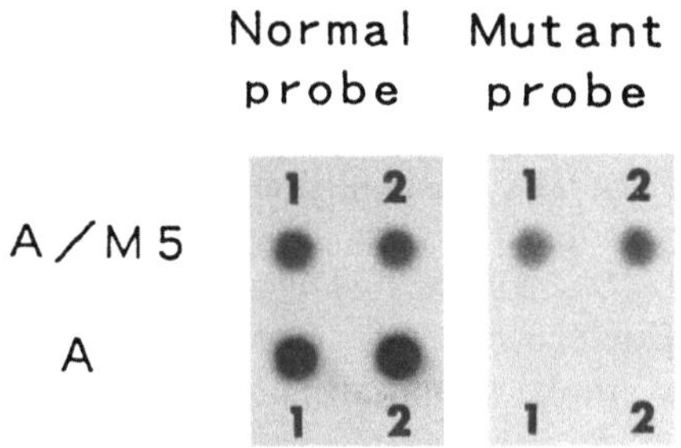

Fig. 4. Dot blot hybridization of amplified DNA of exon 15 of the plasminogen gene derived from subjects with plasminogen phenotypes A and A/M5. The amplified DNA from the subject with plasminogen A/M5 hybridizes with both the normal (A) and mutant (Tochigi) probes

codon 601 missense mutation (GCT → ACT, Ala → Thr). Figure 4 shows representative data of the dot blot hybridization with allele-specific oligonucleotide probe. The results suggest that plasminogen M5 is identical with plasminogen Tochigi. Whether plasminogen M5 (Tochigi) is involved in CHD has not yet been examined. Since the role of the plasminogen-like apo(a) component of Lp(a) in the acute development of thrombi may be one of the causes of the documented association between increased levels of Lp(a) and CHD (Utermann 1989; Scanu and Fless 1990), the possibility exists that plasminogen M5 is also involved in CHD.

Apolipoproteins E-2 (158 Arg → Cys) and E-4 (112 Cys → Arg)

ApoE ia a protein constituent of plasma lipoproteins and plays an important role in lipoprotein metabolism. The genetic polymorphism of apoE is under the control of three common alleles (ε2, ε3, ε4) that specify isoforms apolipoprotein E-2, E-3 and E-4 (apo E-2, apo E-3 and apo E-4), respectively. The molecular basis for apoE polymorphism has been established by Mahley and coworkers. Apo E-2, apo E-3 and apo E-4 differ from one another by single amino acid substitutions, as shown in Fig. 5. Since the ε3 allele is by far the most common in populations, apo E-3 (112 Cys, 158 Arg) is considered to be the parent form of apoE. While the primary molecular defect in type III hyperlipoproteinemia is

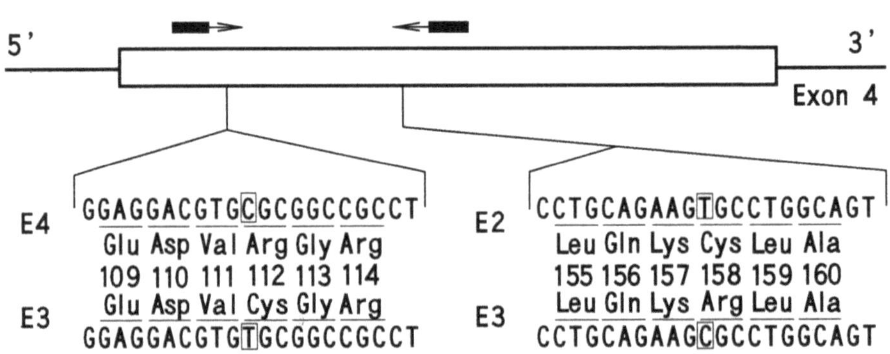

Fig. 5. Partial DNA and amino acid sequences of apo E-3 (112 Cys, 158 Arg), E-2 (112 Cys, 158 Cys), and E-4 (112 Arg, 158 Arg). The positions of primers for PCR and dot blot hybridization with allele-specific oligonucleotide probes are shown by the *solid bars*

the homozygosity of apo E-2 (158 Arg → Cys) in most cases, apo E-2 is usually associated with lower cholesterol levels. There is a high incidence of premature CHD and peripheral atherosclerosis in subjects with type III hyperlipoproteinemia. On the other hand, apo E-4 (112 Cys → Arg) is associated with higher plasma cholesterol levels, though studies of patients with CHD have yielded conflicting information on whether the presence of apo E-4 may impose some risk for developing CHD (Mahley and Rall 1989; Smit et al. 1990).

Using two-dimensional gel electrophoresis, we examined apoE phenotypes of 608 apparently healthy Japanese male adults who visited a hospital in Tokyo for their medical chekups. The subjects were not excluded by any other criterion except diabetes mellitus, abnormal thyroid function or abnormal kidney function. Figure 6 shows the phenotypes of apo E-2/3, apo E-3/3 and apo E-3/4 detectable with two-dimensional gel electrophoresis (Yanagi et al. 1990). Table 1 shows the results of our study together with a summary of the reported phenotype and gene frequencies of apoE in different countries where 400 or more subjects were analyzed. The frequency of the ε2 allele in Japanese is similar to that in the population in Finland, but lower than the frequencies of the ε2 allele in populations in the U.S.A., France, Germany, Scotland, New Zealand, and the Netherlands (Utermann et al. 1982; Menzel et al. 1983; Wardell et al. 1982; Cumming and Robertson 1984; Ehnholm et al. 1986; Lehtimäki et al. 1990; Ordovas et al. 1987; Boerwinkle and Sing 1987; Smit et al. 1988; Eto et al. 1986). In this respect, it is intereting that type III hyperlipoproteinemia is less common in the Japanese than in Caucasians. On the other hand, the frequency of the ε4 allele in the Japanese population is much lower than that in the population in Finland and also seems to be lower than that in populations of the U.S.A. and other European countries.

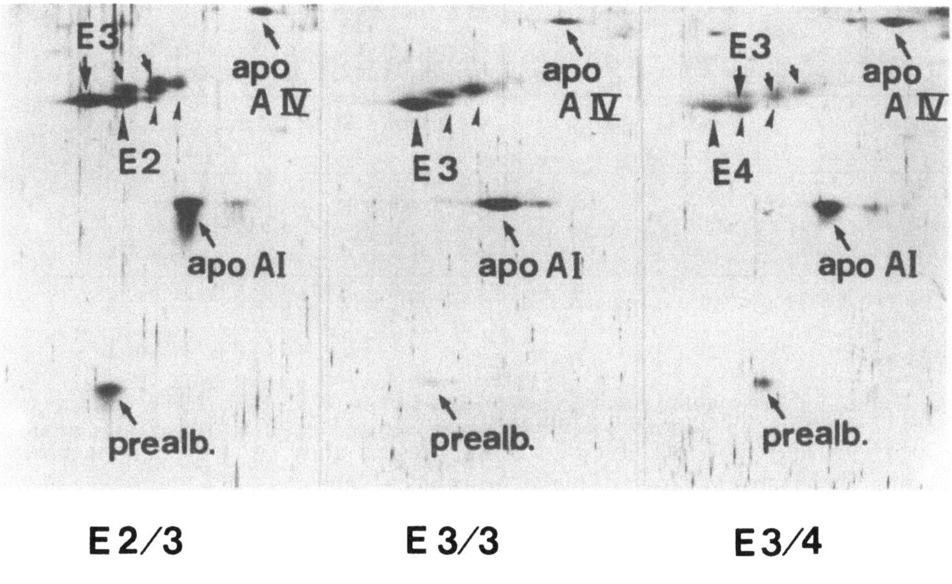

Fig. 6. Three common phenotypes of apoE (E-3/3, E-3/4, E-2/3) in two-dimensional gel electrophoresis patterns

Table 1. Apolipoprotein E phenotypes and gene frequencies in populations

Country	Germany		New Zealand	Scot- land	Finland		USA	France	Nether- lands	Japan	
No. of subjects	1031	1000	462	400	615	1577	1209	434	2018	576	608
Year	1982	1983	1982	1984	1986	1990	1987	1987	1988	1986	1990
Phenotypes					%						
E-3/3	59.8	62.7	51.4	58.3	54.0	58.7	63.3	55.2[a]	55.9	71.9	71.1
E-3/4	22.9	20.2	25.1	24.8	31.9	30.6	18.8	17.5	25.4	19.3	15.8
E-2/3	12.0	11.0	20.0	12.8	6.7	5.4	11.8	20.6	12.9	6.1	8.2
E-2/4	1.5	3.0	1.2	2.8	0.8	1.8	2.0	3.6	2.2	0.7	1.5
E-4/4	2.8	2.3	0.9	1.0	6.3	3.2	3.1	2.2	2.9	1.7	0.7
E-2/2	1.0	0.8	1.4	0.5	0.3	0.6	0.6	0.9	0.7	0.3	0
Alleles					%						
$\varepsilon 3$	77.3	78.3	73.9	77.0	73.3	76.7	78.6	75.1	75.0	84.6	84.1
$\varepsilon 4$	15.0	13.9	14.1	15.0	22.7	19.4	13.5	12.6	16.7	11.7	9.5
$\varepsilon 2$	7.7	7.8	12.0	8.0	4.1	3.9	7.5	12.3	8.2	3.7	4.9
Reference	1	2	3	4	5	6	7	8	9	10	11

Phenotype and allele frequencies of rare apoE variants are omitted.
[a] Phenotype frequencies in France were cited from Boerwinkle (1987).
Reference Key: 1, Utermann et al. (1982); 2, Menzel et al. (1983); 3, Wardell et al. (1982); 4, Cumming and Robertson (1984); 5, Ehnholm et al. (1986); 6, Lehtinäki et al. (1990); 7, Ordovas et al. (1987); 8, Boerwinkle and Sing (1987); 9, Smit et al. (1988); 10, Eto et al. (1986); 11, Present study

Table 2. Mean plasma cholesterol levels in apo E3/3, 3/4 and 2/3 in Japanese male adults

Phenotype	No. of subjects	Age (mean ± SD)	Plasma cholesterol levels (mean ± SD)
E-3/3	432	48.0 ± 7.8	189.6 ± 31.4
E-3/4	96	46.5 ± 8.1	200.7 ± 32.0*
E-2/3	50	46.9 ± 8.3	181.5 ± 32.5**

* $p < 0.005$, ** $p < 0.05$, as compared with cholesterol levels in E-3/3

Table 2 shows mean serum cholesterol levels in subjects with apo E-2/3, apo E-3/3, and apo E-3/4 observed in the present study. As expected, mean serum cholesterol levels were higher in subjects with apo E-3/4 and lower in subjects with apo E-2/3 as compared with the cholesterol level in subjects with apo E-3/3.

It is important to clarify whether the amino acid substitution in apo E-2 and apo E-4 in the Japanese population is identical with that in Caucasians. We have examined this issue using PCR and dot blot hybridization with allele-specific oligonucleotide probes. The sequence of apoE analyzed is shown in Fig. 5. Our preliminary analysis of ten ε2 alleles derived from unrelated Japanese revealed that nine of them were the codon 158 missense mutation (CGC → TGC, Arg → Cys). As to apo E-4, we have so far analyzed 25 ε4 alleles derived from unrelated Japanese and found that all of them are the codon 112 missense mutation (TGC → CGC, Cys → Arg). As in Caucasians, many of apo E-2 are likely to have the codon 158 missense mutation (Arg → Cys) and most, if not all, of apo E-4 have the codon 112 missense mutation (Cys → Arg) in the Japanese population.

Familial Hypercholesterolemia

The level of cholesterol in human plasma to a large extent is controlled by the interaction of LDL and LDL receptor. Defects of the LDL receptor gene cause FH, which is characterized clinically by lifelong elevation in the concentration of LDL in plasma, tendon xanthomas, and premature CHD (Goldstein and Brown 1989). The heterozygote frequency of FH is estimated to be 1 in 500 persons in Japanese as in most Caucasian populations (Mabuchi et al. 1978, 1979). It has been thought that tendon xanthomas are virtually specific for FH. Therefore, when adults with relatively severe hypercholesterolemia associated with tendon xanthomas have hypercholesterolemic first-degree relative(s), they tend to be diagnosed as FH based on clinical findings. However, this supposition has not yet been fully examined. To assess the relationship between relatively severe hereditary hypercholesterolemia with Achilles tendon xanthomas and the defect of the LDL receptor gene, we have carried out pedigree analyses of hypercholesterolemic families (Yamakawa et al. 1988, 1989, 1991; Yuzawa et al. 1989).

The LDL receptor gene was studied in a total of 197 members of 37 Japanese pedigrees, including 17 hypercholesterolemic families with probands having relatively severe hypercholesterolemia (plasma cholesterol levels 300–450 mg/dl) and Achilles tendon xanthomas (above 11 mm thickness in radiography). In the hypercholesterolemic families, members were diagnosed as having primary hypercholesterolemia when plasma cholesterol levels were more than 260 mg/dl in adults and more than 230 mg/dl in children, and when the hypercholesterolemia was neither secondary nor due to high HDL cholesterol levels. The plasma cholesterol level was higher than 300 mg/dl in at least one of the first-degree relatives in 16 of the 17 probands and more than 250 mg/dl in a child of the other one proband who had premature CHD associated with a 22 mm thick Achilles tendon. The LDL receptor gene was analyzed by Southern blot hybridization. We also have used restriction fragment length polymorphisms (RFLPs) of the LDL receptor gene for the family studies. Figure 7 shows the sites of seven RFLPs observed with high polymorphism information content values and 22 RFLP haplotypes detected among 121 normal LDL receptor genes by family studies of 37 Japanese pedigrees.

In 16 of the 17 families, hypercholesterolemia was co-segregated with gross rearrangement, RFLP, or an abnormal TaqI band of the LDL receptor gene.

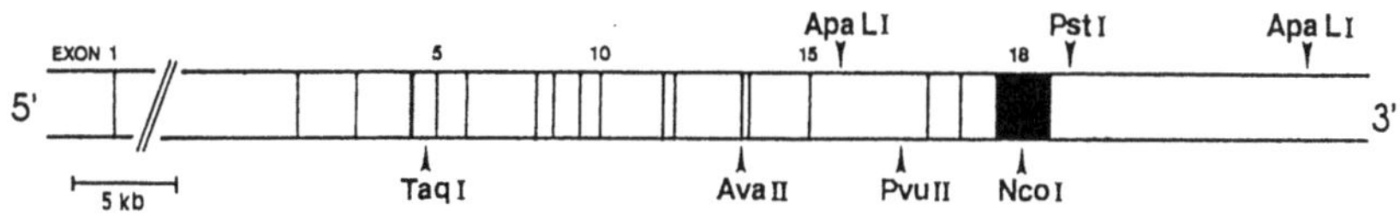

Haplotype	TaqI	AvaII	ApaLI5'	PvuII	NcoI	PstI	ApaLI3'	n (%)
1	–	–	–	–	–	–	–	29 (24.0)
2	+	–	+	–	+	–	+	28 (23.1)
3	+	–	–	–	–	–	–	14 (11.6)
4	+	–	–	–	+	–	+	8 (6.6)
5	–	+	+	–	+	+	+	8 (6.6)
6	–	–	–	+	+	–	+	4 (3.3)
7	+	–	+	–	+	–	–	4 (3.3)
8	–	+	–	–	–	–	–	4 (3.3)
9	–	–	+	–	–	–	–	3 (2.5)
10	+	–	+	–	+	+	+	3 (2.5)
11	–	–	+	–	+	–	+	3 (2.5)
12	–	+	–	+	+	–	+	2 (1.7)
13	–	–	+	–	+	+	+	2 (1.7)
14	–	–	–	–	–	+	–	1 (0.8)
15	–	–	–	–	–	–	+	1 (0.8)
16	–	–	–	+	+	–	–	1 (0.8)
17	+	+	–	–	–	–	–	1 (0.8)
18	+	+	–	–	–	–	+	1 (0.8)
19	+	–	+	–	–	–	+	1 (0.8)
20	–	+	–	–	+	–	+	1 (0.8)
21	+	–	–	+	+	–	+	1 (0.8)
22	+	+	–	+	+	–	+	1 (0.8)
							total	121

Fig. 7. Seven RFLP sites and 22 RFLP haplotypes in the LDL receptor gene detectable in the Japanese population. The map positions of the polymorphic restriction sites are indicated by the *arrowhead*. The presence or absence of each of the polymorphic restriction sites for each of the haplotypes is shown as + or – symbols

These findings are compatible with the interpretation that hypercholesterolemia is due to defective LDL receptor genes. Four representative families are shown in Fig. 8 and 9. In the 16 pedigrees, co-segregated genetic markers of the LDL receptor gene included four different deletion mutations, six different RFLP haplotypes and one TaqI abnormal band, suggesting that the origin of the mutant LDL receptor genes in the Japanese population generally differs among different pedigrees. In the one other family, both the proband and a sibling had relatively severe hypercholesterolemia and Achilles tendon xanthomas and shared an RFLP haplotype; however, the proband's other siblings, with moderate hypercholesterolemia but without Achilles tendon xanthomas, did not. The mutant gene for familial defective apolipoprotein B-100 (3500 Arg → Gln; Soria et al. 1989) was not detected in the 17 probands. These data suggest that most, if not all, of the relatively severe hereditary hypercholesterolemia associated with Achilles tendon xanthomas is caused by a defect of the LDL receptor gene in the Japanese. The analysis of cholesterol levels, Achilles tendon thickness and the LDL receptor gene of family members of probands with relatively severe hypercholesterolemia associated with tendon xanthomas is useful for early diagnosis of FH. Achilles tendon xanthomas can be easily detected by an ultrasonographic method (Yuzawa et al. 1989).

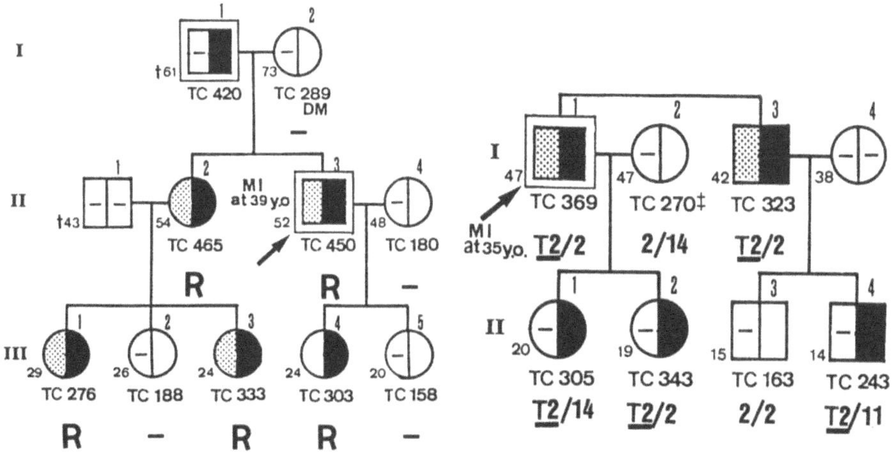

Fig. 8. Pedigrees of two hypercholesterolemic families with probands having relatively severe hypercholesterolemia associated with Achilles tendon xanthomas. *Left*, a pedigree in which a gross rearrangement was observed in the LDL receptor gene. The presence or absence of gross rearrangement is shown as R or –. Deceased members of the family are shown by ± *DM* indicates diabetes mellitus. *Right*, a pedigree in which hypercholesterolemia was co-segregated with an abnormal TaqI band of the LDL receptor gene. *T* indicates the presence of abnormal TaqI band. RFLP haplotypes are shown by large *Arabic numerals*. The proband's wife (I-2) indicated by ‡ had a hight HDL cholesterol level (109 mg/dl). ◨ ◑, primary hypercholesterolemia; ■ ■, Achilles tendon xanthomas; ■ ■, CHD; *TC*, totalcholesterol; *MI*, myocardial infarction

Hereditary Moderate Hypercholesterolemia

To perform genetic studies on hereditary hypercholesterolemia in a population, 2469 Japanese school children ranging in age from 10 to 14 years were first checked for their plasma cholesterol levels before lunch as the first screening. One month later, measurements of fasting serum lipids and routine medical examinations for blood and urine were performed as the second screening in those children who had plasma cholesterol concentrations above 200 mg/dl in the first screening. When plasma cholesterol levels were also above 200 mg/dl in the second screening, the children were diagnosed as having hypercholesterolemia. We diagnoses 201 children (8.2%) of 191 couples of parents as having hypercholesterolemia. Of these parents, 150 couples (both parents in 89 couples and one parent in 61 couples) with hypercholesterolemic children consented to their own examinations for fasting plasma. They included 102 fathers with a mean age of 39 ± 6 and 137 mothers with a mean age of 38 ± 4. When serum cholesterol levels of the parents were over 95 percentile cholesterol levels of controls, they were regarded as having hypercholesterolemia. Hypercholesterolemia was detected in 14% of fathers and 16% of mothers. In families in which at least one parent was hypercholesterolemic, we made efforts to examine fasting plasma lipids in the first-degree relative(s) of the hypercholesterolemic parent. In this study, hypercholesterolemia was regarded as hereditary hyper-

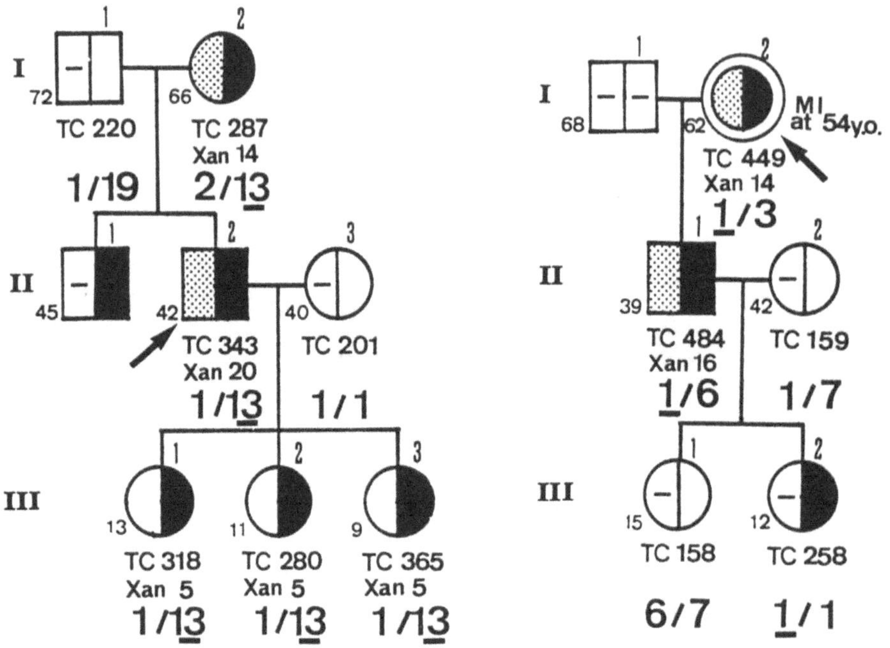

Fig. 9. Pedigrees of two hypercholesterolemic families with probands having relatively severe hypercholesterolemia associated with Achilles tendon xanthomas. The *underlined numerals* indicate the RFLP haplotype of the LDL receptor gene co-segregated with hypercholesterolemia. Marks are the same as in Fig. 8

cholesterolemia whenever primary hypercholesterolemia was observed in the parent of the proband and in one or more first-degree relatives of the hypercholesterolemic parent, excepting the proband. Although numbers of families in which first-degree relatives of hypercholesterolemic parents could be examined were not large, we have detected 14 pedigrees with hereditary hypercholesterolemia (Table 3).

Table 3. Results of a population and family study on hereditary hypercholesterolemia

Number of school children examined	2469
Number of hypercholesterolemic children	201
Number of families examined	150
Both parents examined	89
One parent examined	61
Number of pedigrees with hereditary hypercholesterolemia detected	14
Classic familial hypercholesterolemia	2
Possible classic familial hypercholesterolemia	1
Moderate hypercholesterolemia	11
Co-segregated with LDL receptor RFLPs	1
Not linked with LDL receptor RFLPs	5
Not informative	1
Not analyzed	4

Among these 14 pedigrees, two were diagnosed as classic FH and one was diagnosed as possible classic FH, based on clinical findings. Hypercholesterolemia in the remaining 11 pedigrees is moderate (250–300 mg/dl in adults) and no hypercholesterolemic adults had Achilles tendon xanthomas. The mutant apoB gene for familial defective apolipoprotein B-100 (3500 Arg → Gln; Soria et al. 1989) was not detected in any proband of the 14 pedigrees. We have analyzed a linkage relationship between the LDL receptor gene RFLPs and hypercholesterolemia in family members of seven of the 11 pedigrees with moderate hypercholesterolemia. We could not get enough samples necessary for the linkage analysis in the other four pedigrees. No linkage relationship was observed in five of the seven pedigrees (Fig. 10). Hypercholesterolemia and the LDL receptor gene RFLPs were co-segregated in one of the seven pedigrees. The linkage relationship was not informative in the remaining pedigree (Table 3).

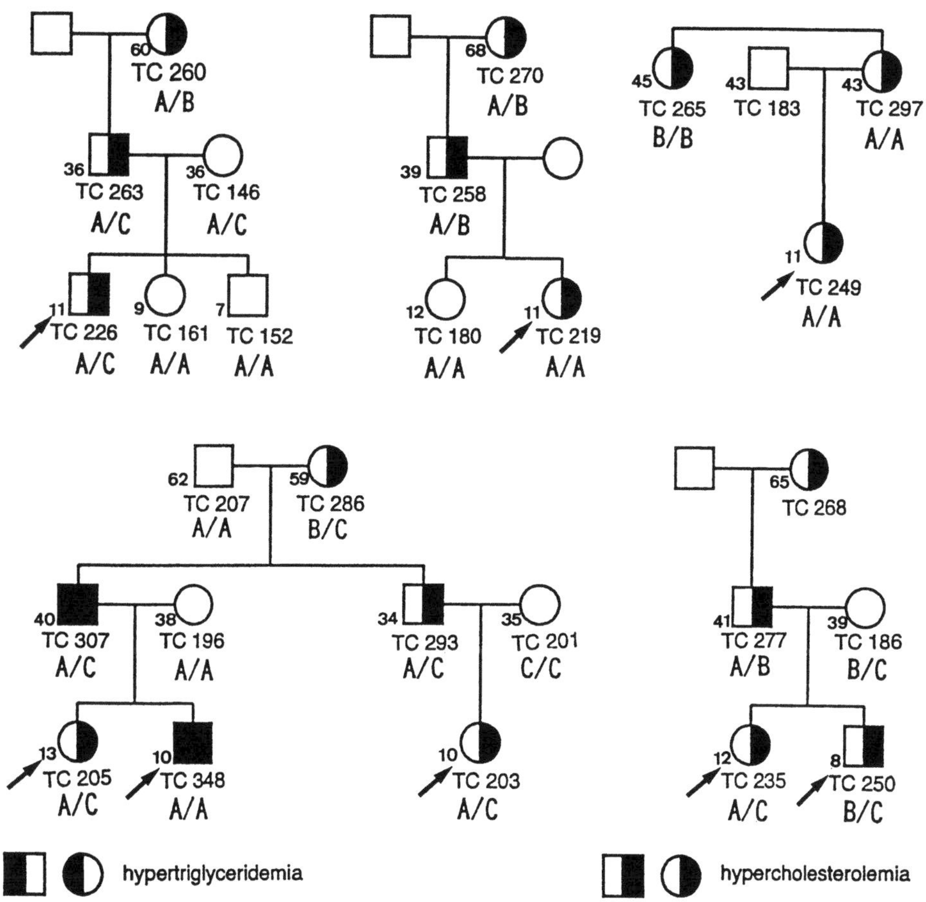

Fig. 10. Five pedigrees with moderate hypercholesterolemia in which co-segregation of hypercholesterolemia and the LDL receptor gene RFLP haplotype was not observed. RFLP haplotypes of the LDL receptor gene are symbolically indicated by *A*, *B* and *C*. Total cholesterol (*TC*) levels and ages are shown

These data suggest that hereditary moderate hypercholesterolemia due to abnormal gene(s) distinct from the abnormal LDL receptor gene may be more common than classic FH in the Japanese population. Since the probands are hypercholesterolemic children, this type of hereditary hypercholesterolemia is considered to manifest itself from childhood. It is likely that the hereditary moderate hypercholesterolemia described here is monogenic in some pedigrees and oligogenic in pedigrees.

Molecular genetic studies of hereditary moderate hypercholesterolemia are now in progress.

Apolipoprotein A-I Deficiency and Premature CHD Due to a Nonsense Mutation in the Apolipoprotein A-I Gene

Human apoA-I is a single polypeptide chain composed of 243 amino acids and plays a central role in HDL metabolism as the major protein of HDL and as a cofactor for lecithin : cholesterol acyltransferse. Decreased plasma apoA-I, as well as a low HDL cholesterol level, has been shown to be a risk factor for CHD (Breslow 1989). It has been suggested that half of the HDL cholesterol variability is genetic and the ther half environmental. The genetic component might be due to genetic variation at a few loci that control HDL metabolism. The apolipoprotein gene locus on chromosome 11, which includes the apoA-I gene, has been postulated as one candidate locus (Breslow 1989). To examine whether a mutant gene at the apoA-I gene locus can cause apoA-I deficiency and a marked reduction of HDL cholesterol, we have analyzed the apoA-I gene of a Japanese female patient with apoA-I deficiency and premature CHD.

Four pedigrees of patients with apoA-I deficiency and very low HDL cholesterol levels have been described (Gustafson et al. 1979; Norum et al. 1982; Schaefer et al. 1985; Hiasa et al. 1986; Karathanasis et al. 1987; Ordovas et al. 1989). The patients all suffered from premature CHD, supporting the concept of a role for apoA-I in the pathogenesis of atherosclerosis. Furthermore, pedigrees of two patients with apoA-I deficiency due to a frameshift mutation in the apoA-I gene have been described in a review article (Assmann et al. 1990). One of the patients with apoA-I deficiency described above is a Japanese female, an offspring from a first cousin marriage (Fig. 11; Hiasa et al. 1986). We analyzed the apoA-I and the apoA-I gene of this Japanese patient.

The patient's plasma density fraction from 1.063 to 1.21 g/ml contained no apoA-I on SDS-polyacrylamide gel electrophoresis and no measurable HDL cholesterol. Southern blot hybridization showed that no gross abnormality is present in the patient's apoA-I gene and that she is homozygous for a RFLP haplotype T2X1M1P1S1 in the apoA-I gene region. Sequencing after amplification by PCR revealed the presence of a codon 84 nonsense mutation (CAG $\rightarrow$ TAG, Gln $\rightarrow$ Stop) of exon 4 in the patient's apoA-I gene (Fig. 12). The data from dot blot hybridization with allele-specific oligonucleotide probes indicated that she is homozygous for the apoA-I gene with regard to the nonsense mutation. These findings indicate that homozygosity for the apoA-I gene with

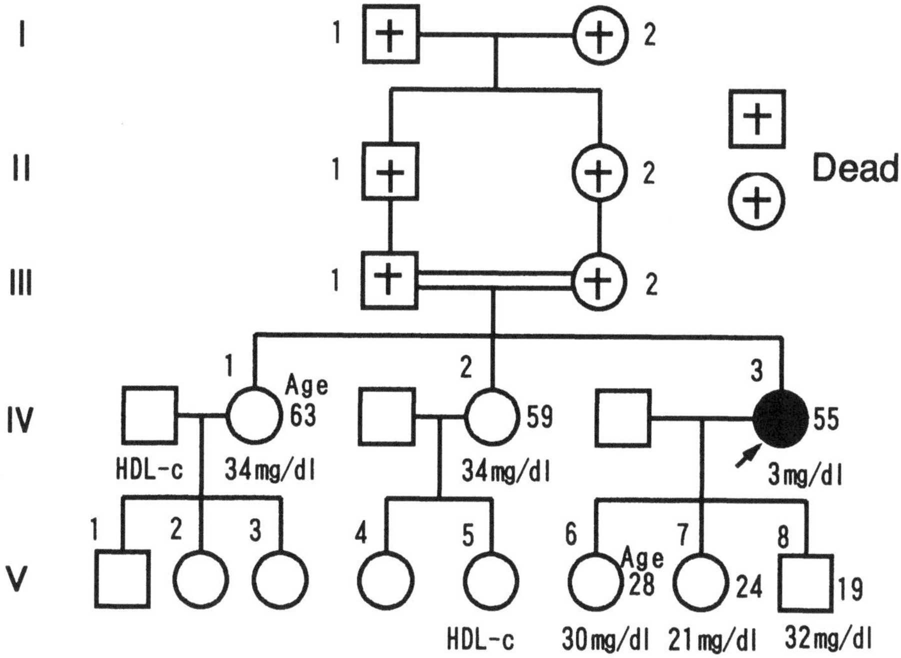

Fig. 11. The pedigree of a kindred with apoA-I deficiency. The patient, indicated by the *arrow*, was noted to have planar xanthomas at age 18, and her symptom of angina pectoris started at age 52. She was diagnosed as having myocardial infarction at age 55. HDL cholesterol (*HDL-c*) levels of the patient's sisters and children are shown

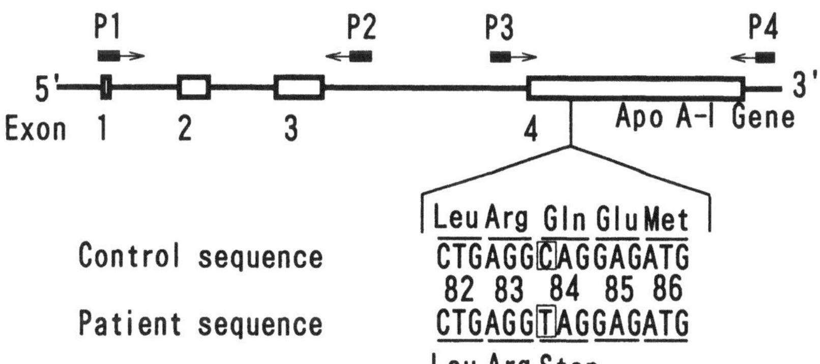

Fig. 12. The patient and normal sequences in the apoA-I gene. The C to T substitution in codon 84 of exon 4 is highlighted by the *box*. The positions of PCR are shown (*P1–P4*)

codon 84 nonsense mutation causes the deficiency of apoA-I and HDL cholesterol in the patient (Matsunaga et al., submitted).

The result of this study suggest that a mutant apoA-I gene can act as a major gene for decreased levels of apoA-I and HDL cholesterol. Indeed the patient's children were reported to have decreased levels of HDL cholesterol (Fig. 11). They are obligate heterozygotes for the mutant apoA-I gene with the codon 84 nonsense mutation. Their decreased HDL cholesterol levels were probably caused by the mutant apoA-I gene.

Applications to the Prevention of Premature CHD

When genotypes strongly associated with premature CHD are elucidated, it is possible to apply the analysis of the genotypes to the prevention of CHD. Thus the genetic analyses of the LDL receptor gene and apoE are now applied for the early diagnosis and treatment of FH and type III hyperlipoproteinemia. The analysis of apo(a) isoforms, combined with the meaurement of Lp(a) levels, is also applicable to the prevention of premature CHD.

In the case of applications of genetic data for the prevention of premature CHD, it is very useful to carry out examinations on family members of probands, because members with genetic risk factor(s) for CHD are efficiently detectable. At present, examinations of family members of patients with FH are especially useful for the early diagnosis and treatment of FH.

For genetic approaches to the prevention of premature CHD, it is important to elucidate genotypes at plural loci which are strongly associated with CHD. In this respect, polymorphic genes associated with CHD or possibly involved in CHD are important, because these genes are very common and might be to a greater or lesser degree involved in CHD in many patients, in combination with other genetic factors. Further association studies with CHD are needed for apo(a) isoforms with smaller molecular weights, plasminogen M5, and apo E-4, in combination with other genetic factors. In the case of genetic variability of lipoprotein metabolism, ist is also important to elucidate new loci or new polymorphic alleles, or both, which are responsible for increased in LDL cholesterol levels or decreases in HDL cholesterol levels.

References

1. Assman G, Schmitz G, Funke H, von Eckardstein A (1990) Apolipoprotein A-I and HDL deficiency. Curr Opin Lipidol 1: 110–115
2. Boerwinkle E, Sing CF (1987) The use of measured genotype information in the analysis of quantitative phenotypes in man. Ann Hum Genet 51: 211–226
3. Boerwinkle E, Visvikis S, Welsh D, Steinmetz J, Hanash SM, Sing CF (1987) The use of measured genotype information in the analysis of quantitative phenotype in man. II. The role of the apolipoprotein E polymorphism in determining levels, variability, and covariability of cholesterol, betalipoprotein, and triglycerides in a sample of unrelated individuals. Am J Med Genet 27: 567–582
4. Breslow JL (1989) Familial disorders of high density lipoprotein metabolism. In: Scriver CR, Beaudet AL, Sly WS, Valle D (eds) The metabolic basis of inherited disease, 6th edn. McGraw-Hill, New York, pp 1251–1266
5. Cumming AM, Robertson FW (1984) Polymorphism at the apoprotein-E locus in relation ro risk of coronary disease. Clin Genet 25: 310–313
6. Ehnholm C, Lukka M, Kuusi T, Nikkila E, Utermann G (1986) Apolipoprotein E polymorphism in the Finish population: gene frequencies and relation to lipoprotein concentrations. J Lipid Res 27: 227–235

7. Eto M, Watanabe K, Ishii K (1986) A racial difference in apolipoprotein E allele frequencies between the Japanese and Caucasian populations. Clin Genet 30: 422–427
8. Gaubatz JW, Ghanem KI, Guevera J, Nava ML, Patsch W, Morrisett JD (1990) Polymorphic forms of human apolipoprotein(a): inheritance and relationship of their molecular weights to plasma levels of lipoprotein (a). J Lipid Res 31: 603–613
9. Goldstein JL, Brown MS (1989) Familial hypercholesterolemia. In: Scriver CR, Beaudet AL, Sly WS, Valle D (eds) The metabolic basis of inherited disease, 6th edition. McGraw-Hill, New York, pp 1215–1250
10. Gustafson A, McConathy WJ, Alaupovic P, Persson B (1979) Identification of lipoprotein families in a variant of human plasma apolipoprotein A deficiency. Scand J Clin Invest 39: 377–387
11. Hiasa H, Maeda T, Mori H (1986) Deficiency of apolipoproteins A-I and C-III and severe coronary heart disease. Clin Cardiol 9: 349–352
12. Karathanasis SK, Ferris E, Haddad IA (1987) DNA inversion within the apolipoprotein AI/CIII/AIV-encoding gene cluster of certain patients with premature atherosclerosis. Proc Natl Acad Sci USA 84: 7198–7202
13. Lehtimaki T, Moilanen T, Viikari J, Akerblom HK, Ehnholm C, Ronnemaa T, Marniemi J, Dahlen G, Nikkari T (1990) Apolipoprotein E phenotypes in Finnish youths: a cross-sectional and 6-year follow-up study. J Lipid Res 31: 487–495
14. Mabuchi H, Tatami R, Haba T, Ueda K, Ueda R, Kametani T, Itoh S, Koizumi J, Ohta M, Miyamoto S, Takeda R, Takeshita H (1978) Homozygous familial hypercholeterolemia in Japan. Am J Med 65: 290–297
15. Mabuchi H, Tatami R, Ueda K, Ueda R, Haba T, Kametani T, Watanabe A, Wakasugi T, Ito S, Koizumi J, Ohta M, Miyamoto S, Takeda R (1979) Serum lipid and lipoprotein levels in Japanese patients with familial hypercholesterolemia. Atherosclerosis 32: 435–444
16. Mahley RW, Rall SC (1989) Type III hyperlipoproteinemia (Dysbetalipoproteinemia). In: Scriver CR, Beaudet AL, Sly WS, Valle D (eds) The metabolic basis of inherited disease, 6th edn. McGraw-Hill, Yew York, pp 1195–1213
17. Menzel HJ, Kladetzky RG, Assmann G (1983) Apolipoprotein E polymorphism and coronary artery disease. Arteriosclerosis 3: 310–315
18. Miyata T, Iwanaga S, Sakata Y, Aoki N (1982) Plasminogen Tochigi: Inactive plasmin resulting from replacement of alanine-600 by threonine in the active site. Proc Natl Acad Sci USA 79: 6132–6136
19. Norum RA, Lakier JB, Goldstein S, Angel A, Goldberg RB, Block WD, Nofze DK, Dolphin PJ, Edelglass J, Bogorad DD, Alaupovic P (1982) Familial deficiency of apolipoprotein A-I and C-III and precocious coronary-artery disease. N Engl J Med 306: 1513–1519
20. Ordovas JM, Litwack-Klein L, Wilson PWF, Schaefer MM, Schaefer EJ (1987) Apolipoprotein E isoform phenotyping methodology and population frequency with identification of apoE1 and apoE5 isoforms. J Lipid Res 28: 371–380
21. Ordovas JM, Cassidy DK, Civcira F, Bisgaier CL, Schaefer EJ (1989) Familial apolipoprotein A-I, C-III, and A-IV deficiency and premature atherosclerosis due to deletion of a gene complex on chromosome 11. J Biol Chem 264: 16339–16342
22. Petersen TE, Martzen MR, Ichinose A, Davie EW (1990) Characterization of the gene for human plasminogen, a key proenzyme in the fibrinolytic system. J Biol Chem 265: 6104–6111
23. Scanu AM, Fless GM (1990) Lipoprotein (a). J Clin Invest 85: 11709–1715
24. Schaefer EJ, Ordovas JM, Law SW, Ghiselli G, Kashyap ML, Srivastava LS, Heaton WH, Albers JJ, Connor WE, Lindgren FT, Lemeshev Y, Segrest JP, Brewer HB (1985) Familial apolipoprotein A-I and C-III deficiency, variant II. J Lipid Res 26: 1089–1101
25. Smit M, de Knijff P, Rosseneu M, Bury J, Klasen E, Frants R, Havekes L (1988) Apolipoprotein E polymorphism in the Netherlands and its effect on plasma lipid and apolipoprotein levels. Human Genet 80: 287–292
26. Smit M, de Knijff P, van der Kooij-Meijs E, Groenendijk C, van den Maagdenberg AMJM, Leuven JAG, Stalenhoef AFH, Stuyt PMJ, Frants RR, Havekes LM (1990) Genetic heterogeneity in familial dysbetalipoproteinemia. The E2(Lys146ü|Gln) variant results in a dominant mode of inheritance. J Lipid Res 31: 45–53

27. Soria LF, Ludwig EH, Clarke HRG, Vega GL, Grundy SM, McCarthy BJ (1989) Association between a specific apolipoprotein B mutation and familial defective apolipoprotein B-100. Proc Natl Acad Sci USA 86: 587–591
28. Utermann G, Steinmets A, Weber W (1982) Genetic control of human apolipoprotein E polymorphism: Comparison of one- and two-dimensional techniques of isoprotein analysis. Human Genet 60: 344–351
29. Utermann G (1989) The mysteries of lipoprotein (a). Science 246: 904–910
30. Yamaguchi M, Doi S, Yoshimura M (1989) Plasminogen phenotypes in a Japanese population. Human Hered 39: 356–360
31. Yamakawa K, Okafuji T, Iwamura Y, Yuzawa K, Satoh J, Hattori N, Yamanouchi Y, Yanagi H, Kawai K, Tsuchiya S, Russell DW, Hamaguchi H (1988) TaqI polymorphism in the LDL receptor gene and a TaqI 1.5-kb band associated with familial hypercholesterolemia. Human Genet 80: 1–5
32. Yamakawa K, Takada K, Yanagi H, Tsuchiya S, Kawai K, Nakagawa S, Hamaguchi H (1989) Three novel partial deletions of the low-density lipoprotein gene in familial hypercholesterolemia. 82: 317–321
33. Yamakawa K, Yanagi H, Saku K, Sasaki J, Okafuji T, Shimakura Y, Kawai K, Tsuchiya S, Takada K, Naito S, Arakawa K, Hamaguchi H. Family studies of the LDL receptor gene of relatively severe hereditary hypercholesterolemia associated with Achilles tendon xanthomas. Human Genet in press
34. Janagi H, Shimakura Y, Yamanouchi Y, Watanabe Y, Tsuchiya S, Hamaguchi H (1990) Association of hypercholesterolemia and apolipoprotein E4 in school children. Clin Genet 38: 264–269
35. Yuzawa K, Yamakawa K, Tohno E, Seki M, Akisada M, Yanagi H, Okafuji T, Yamanouchi Y, Hattori N, Kawai K, Shimakura Y, Tsuchiya S, Ijima H, Fukao K, Iwasaki Y, Hamaguchi H (1989) An ultrasonographic method for detection of Achilles tendon xanthomas in familial hypercholesterolemia. Atherosclerosis 75: 211–218
36. Wardell MR, Suckling PA, Janus ED (1982) Genetic variation in human apolipoprotein E. J Lipid Res 23: 1174–1182

Genetic Epidemiology of Ischemic Heart Disease and Arterial Hypertension

V. A. Koshechkin, C. B. Hudaiberdyev, V. A. Kabakova, S. V. Panin, T. A. Rozhkova, and *E. Y. Solovjova*

Cardiovascular diseases give rise to high morbidity, mortality and infirmity, inflicting heavy social losses on a population (Chazov and Klimov 1980; Oganov 1990). Ischaemic heart disease (IHD) and high blood pressure (HBP) represent a major contribution (Chasov et al. 1986; WHO 1985, 1986).

Preventive studies demonstrate that the morbidity and mortality increase caused by IHD and HBP can be stopped. A successful struggle against IHD and HBP largely depends upon improvements in diagnosis, treatment and prevention. It also requires a new knowledge on interaction of IHD and HBP with predisposing factors (Koshechkin 1983; Koshechkin et al. 1986; Berg 1987; Fredrickson and Levy 1972; Williams et al. 1988). Defining the role of genetic factors in IHD and HBP in certain populations is another area to investigate.

The aim of this chapter is to examine the role of genetic and environmental factors in the incidence of IHD and HBP in populations of various regions of the USSR.

The following principle underlay the planning and completion of the study: that the interaction of two components is needed for IHD and HBP development,
1. genetic predisposition to pathology and
2. specific environmental component.

The effect of the latter is significant at a certain crucial moment of the organism's life. The greater the importance of the genetic lesions, the lower the environmental effect needed for pathogenetic triggering and vice versa.

When embarking upon the study, we suggested that the interactions of genetic and environmental factors of IHD and HBP differed considerably between various ethnic and geographically distinct populations.

Materials and Methods

The following populations were chosen as subjects for genetic-epidemiological study:
1. 2502 men aged between 20–59 years, living in Ashchabad city, the capital of Turkmenskaya USR.

K. Berg, V. Bulyzhenkov, Y. Christen, P. Corvol (Eds.)
Genetic Approaches to Coronary Heart Disease and Hypertension

2. The population (2162) of Nochur settlement in the Bacharden rural area, in the vicinity of Ashchabad city.
3. 3321 men aged between 20–59 years, living in Moscow.
4. Random sample of families living in Moscow (1431 males and females).

A sample of men aged between 20–59 years living in Ashchabad were examined in the course of a cooperative epidemiologic study performed in accordance with a program called "Epidemiology of IHD and artherosclerosis in various regions of the country". In the first stage (cross-section study), the incidence of the main forms of IHD and their risk factors were studied in a random population of men aged between 20–59 years. The number of those examined in the Ashchabad sample was 3201 men, or 64 % of men picked at random.

The settlement of Nochur is situated in the Bacharden rural area in the civinity of Ashchabad city; its highland inhabitants are of special interest to us, due to their unique population structure and high prevalence of HBP, peculiarities of the population's migration (emigration presumably), high percentage of close-bred marriages, and high inbreeding (no less than 1–1.5 %).

The sample of men examined in Moscow can actually be divided into two samples, formed in two stages.

In the first stage, a random selection of 5000 male inhabitants of a Moscow district aged between 20–59 years were chosen, 3331 (66.6 %) of them were then examined.

In the second stage, probands were selected among the men who had been examined. They displayed either symptoms of IHD, plasma cholesterol concentrations over 260 mg/dl, or triglyceride concentrations over 200 mg/dl, or all of these. Close relatives of probands were examined next. A total of 1431 people were examined in the family sample.

All examinations were performed according to a methodology of epidemiological studies created in the All Union Cardiology Research Centre USSR Academy of Medical Sciences.

The physical examination program included:

1. Standard questioning according to the WHO questionnaire on revealing angina pectoris.
2. ECG registration in 12 standard leads at rest, following encoding by Minnesota Code (MC).
3. Anthropometric examination (height in cm, to within 0.5 cm; body mass in kilos, to within 0.1 kilo). Body mass was appraised by the Ketle Index. Excessive body mass equalled Ketle Index = >29.0.
4. Blood pressure was measured twice, with 5 minute intervals, on the right arm, in a seated position. Mean value for blood pressure was calculated for two measurements; both for systolic and diastolic pressures.
5. Defining total concentrations of cholesterol, triglycerides and HDL cholesterol in blood serum by means of a "Technicon" analyser. The study distinguishes strict and probable criteria of IHD depending upon the results of questioning and ECG.

Strict criteria of IHD included (Rose and Blackburn 1968):
1. The revelation of myocardial infacrtion with Minnesota Code scores 1-1 (all), 1-2-1 (without 1-2-8).
2. Angina pectoris revealed by Rose questioning.
3. Painless form of IHD with Minnesota Code scores 4-1, 5-1, without 3-1 and 3-3.

Probable criteria of IHD included:
1. Minnesota Code scores 1-2-8, 1-3 (all), 4-3.
2. Minnesota Code scores 4-1 and 5-1, 2 with 3-1 and 3-3, 6-1, 7-1 and 8-3 either, for patients over 40 years of age only.
3. Possible myocardial infarction.

A set of multivariate analysis procedures secured by SAS (Statistical Analysis System 1982) programs was applied to an interaction analysis study of IHD and risk factors; the latter included traits of familial anamnesis. Determination of interaction types of linear dependence was performed by means of the General Linear Model (GLM) procedure; the program classified interaction levels of independent variables and a studied trait, simultaneously employing variables of discreet and continuous levels.

Determination of multiple regression analysis was performed by means of the STEPWISE regression analysis procedure (SAS).

From the multitude of independent variables the STEPWISE procedure selects those that must be included in a regression model with corresponding statistical significance.

Genetic-statistical analysis was performed with application of monolocus diallelic model and breakdown of phenotypical variance of different traits into genetic and environmental components.

Results

Genetic-epidemiological description of IHD in men living in the cities of Moscow and Ashchabad.

The incidence of IHD in men aged between 20–59 years is by no means significantly higher in Ashchabad than in Moscow: 13.8% and 12.7%, respectively (see Table 1).

Both in Ashchabad and in Moscow the risks of IHD development were 2–3 times higher in men whose parents had suffered insult or heart attack or had died from such events than in men whose parents had never suffered these particular diseases or whose cause of death was not related to cardiovascular pathology. At the same time, relative IHD risks were higher in those particular men whose mothers had suffered or died from cardiovascular diseases than whose fathers had been so affected (see Table 2).

This finding is also confirmed by risk factor trait distributions: age, systolic (SBP) and diastolic (DBP) blood pressure, body mass index (BMI), total cholesterol (CH), triglycerides (TG), and high density lipoproteins cholesterol

Table 1. IHD incidence among men aged between 20–59 years living in Ashchabad and Moscow

Age groups (years)	Ashchabad		Moscow		Statistical significance ($p < 0.05$)
	Total number	IHD percentage	Total number	IHD percentage	
20–29	673	3.86	732	3.03	n. s.[a]
30–39	763	10.13	792	5.64	0.001
40–49	623	18.13	938	12.17	0.001
50–59	443	28.90	864	27.91	0.001
20–59	2502	13.80	3321	12.70	0.001

[a] n. s., not significant

Table 2. Relative IHD risk in men aged between 20–59 years living in Moscow and Ashchabad whose parents have died of stroke or heart attack

Age groups (years)	Ashchabad city				Moscow city			
	Cardiovascular diseases revealed in anamnesis of:							
	Mother		Father		Mother		Father	
	% men with IHD	Relative risk	% men with IHD	Relative risk	% men with IHD	Relative risk	% men with IHD	Relative risk
20–29	13.33	3.70	3.03	0.82	1.86	0.83	10.00	2.16
30–49	19.15	2.40	11.39	1.48	5.00	0.77	6.48	1.20
40–49	17.24	0.97	7.25	0.36	16.35	2.12	12.88	1.36
50–59	40.00	2.17	52.50	2.63	27.43	1.36	24.18	2.22

(HDL-CH); bearing in mind that it depends on the presence (or absence) of IHD traits in men aged between 20–59 years living in Moscow and in Ashchabad (see Table 3). It demonstrates that age, SBP, DBP and CH concentrations are higher in persons with IHD traits in both samples studied, whereas HDL-CH concentrations are lower in the same persons.

There are differences in distribution of the data concerning cardiovascular diseases among parents of men aged between 20–59 years living in Ashchabad and Moscow with and without IHD (see Table 4).

Determination of linkage between IHD manifestations and risk factors, including familial anamnesis data was performed by means of the GLM procedure (see Table 5).

Cardiovascular disease indication in familial anamnesis of parents in Ashchabad is significant in IHD pathogenetic prognosis; it also takes into account age, arterial blood pressure, and BMI.

However, the structure of IHD traits in Moscow inhabitants differs considerably from the trait structure of Ashchabad dwellers. For instance, familial

Table 3. Distribution (X ± m) of risk factors in men aged between 20–59 years living in Ashchabad and Moscow with and without IHD

Traits	Ashchabad			Moscow		
	IHD−	IHD+	Statistical significance	IHD−	IHD+	Statistical significance
	X ± m	X ± m		X ± m	X ± m	
SBP	126 ± 0.6	136 ± 1.2	0.0001	131 ± 0.8	141 ± 1.3	0.0001
DBP	83 ± 0.4	88 ± 0.9	0.0001	86 ± 0.2	91 ± 0.7	0.0001
Age	36 ± 0.2	44 ± 0.5	0.0001	39 ± 0.2	49 ± 0.2	0.0001
BMI	24 ± 0.06	28 ± 0.2	0.0001	25 ± 0.05	27 ± 0.1	0.0001
	–	–	–	209 ± 0.8	221 ± 2.1	0.0001
TG	–	–	–	103 ± 1.4	124 ± 3.5	0.0001
HDL-CH	–	–	–	51 ± 0.2	46 ± 0.6	0.0001

Table 4. Distribution (%) of family anamnesis data in men aged between 20–59 years living in Ashchabad and Moscow with and without IHD

Traits	Ashchabad			Moscow		
	IHD−	IHD+	Statistical significance a ($p < 0.05$)	IHD−	IHD+	Statistical significance a ($p < 0.05$)
			Cause of mother's death			
Heart attack	3.0 ± 0.3	5.2 ± 1.2	n. s.	4.4 ± 0.4	10.6 ± 1.5	0.001
Stroke	3.2 ± 0.4	8.3 ± 1.5	0.01	5.1 ± 0.4	8.6 ± 1.4	0.05
			Cause of father's death			
Heart attack	4.0 ± 0.4	9.5 ± 1.6	0.0001	9.2 ± 0.5	10.6 ± 1.5	n. s.
Stroke	3.6 ± 0.4	4.6 ± 1.1	n. s.	3.2 ± 0.4	7.6 ± 1.3	0.01
			Mother's anamnesis			
Heart attack	17.3 ± 0.9	24.0 ± 2.5	0.05	28.1 ± 0.1	33.1 ± 2.6	n. s.
Stroke	5.4 ± 0.5	10.1 ± 1.7	0.01	9.2 ± 0.5	14.6 ± 1.8	0.01
			Father's anamnesis			
Heart attack	12.9 ± 0.8	15.2 ± 2.2	n. s.	25.0 ± 0.9	21.7 ± 2.4	n. s.
Stroke	5.2 ± 0.5	7.4 ± 1.5	n. s.	11.0 ± 0.6	15.0 ± 1.9	0.05

Symbols: a, T-Student criteria; n. s., not significant

anamnesis data contribute less significantly to IHD pathogenetic prognosis in Moscow inhabitants then in men living in Ashchabad. The R-square is the same in both groups and equals 9.07 % and 9.36 %, respectively.

At the same time, every age group has its own peculiarities. For instance, in Ashchabad inhabitants, age is a risk factor only in the 40–49 age group and does not significantly influence other age groups. On the whole, age is one of the major factors of IHD pathogenesis and forms the greatest part of the R-square.

Table 5. Significance of IHD and familial anamnesis data connection in men aged between 20–59 years living in Ashchabad and Moscow

Traits	Ashchabad		Moscow	
	F-value	T-criterial	F-value	T-criteria
Age	169.02	0.0001	263.29	0.0001
Systolic blood pressure	27.31	0.0001	27.07	0.0001
Diastolic blood pressure	1.03	0.3103	0.00	0.9896
Body mass index	12.29	0.0015	40.91	0.0001
Mother's anamnesis				
Heart attack	2.12	0.0941	1.67	0.1884
Stroke	1.91	0.1246	0.01	0.9662
Cause of mother's death				
Heart attack	0.26	0.6111	5.08	0.0242
Stroke	5.67	0.0174	1.10	0.2936
Father's anamnesis				
Heart attack	1.10	0.3473	0.74	0.5815
Stroke	1.36	0.2512	0.54	0.5815
Cause of father's death				
Heart attack	12.58	0.0004	0.08	0.7830
Stroke	0.04	0.8410	0.08	0.7725
Probability of null hypothesis		0.0001	0.0001	
Variance of the true error		9.07	9.36	

The following indexes of familial anamnesis should be included as independent traits in a regression model of IHD probability in men aged between 20–59 years living in Ashchabad:

1. Stroke as cause of mother's death.
2. Father's heart attack.

The regression model of IHD probability in men aged between 20–59 years living in Moscow should include the mother's death by heart attack as an independent trait.

The equation of multiple regression for men aged between 20–59 years living in Ashchabad city is:

$$Y = -0.4628 + 0.006\,(\text{age}) + 0.0015\,(\text{SBP}) + 0.0063\,(\text{BMI}) + 0.1022\,(\text{stroke as cause of mother' death}) + 0.1265\,(\text{father's heart attack}).$$

The equation of multiple regression for men aged between 20–59 years living in Moscow is:

$$Y = -0.4948 + 0.006\,(\text{age}) + 0.0009\,(\text{SBP}) + 0.0097\,(\text{BMI}) + 0.0642\,(\text{stroke as cause of mother's death}).$$

Statistical analysis results were appraised by means of GLM and STEPWISE regression analysis of examination results in men of various age groups living in Moscow and in Ashchabad. Familial anamnesis indications of death due to stroke or heart attack were of even greater significance for IHD prognosis than the proband's own traits, i.e. age, SBP, and BMI.

The interaction of genetic and environmental factors in IHD manifestations in men aged between 20–59 years was analysed via the testing of a monolocus diallelic model. Results indicate that:

1. The hypothesis about the complete dominance of genetic factors which determine IHD incidence cannot be accepted.
2. IHD incidence in a population cannot be determined solely by environmental factors.
3. The monolocus diallelic model describes the examined sample most successfully according to a quasi dominant variant of inheritance with incomplete penetrance of AA and Aa genotypes.

The Contribution of Genetic and Environmental Factors to Concentration Variability of Plasma Lipids

During the examination of family members of men (probands) aged between 20–59 years living in Moscow, a sample of 1431 men and women was created. It included both probands and their close relatives. Because studied traits possess expressed age regressions which are statistically significant, they were assigned to the age of 45 years.

The results of the family sample study enabled us to calculate correlation coefficients for pairs of relatives which later were used for breaking down the phenotypic variance into genetic and environmental components (see Table 6).

The acquired data indicate that interindividual variability of cholesterol, triglyceride and HDL-cholesterol concentrations are largely determined by the combined action of genetic and environmental factors. In the case of cholesterol concentrations, the importance of the dominant component was clarified. It may mean the probability of genes with main effect influence.

Table 6. Breakdown of phenotypical variance of cholesterol, triglycerides and HDL-cholesterol concentrations into genetic and environmental components

Traits	Variance components				
	Autosomal		Systematic environmental		Maternal effect
	Dominant	Additive	One generation	Random	
Cholesterol	28	44	4	16	6
Triglycerides	0	56	0	44	0
HDL-cholesterol	0	41	24	30	5

Epidemiologic Genetics of Dislipoproteinaemias (DLP)

The structure of family dislipoproteinaemias was studied by means of distribution analyses of cholesterol, triglyceride and HDL-cholesterol concentrations in men aged between 20–59 years living in a Moscow district.

Concentrations of cholesterol over 260 mg/ml, triglycerides over 200 mg/ml and HDL-cholesterol lower than 35 mg/ml were accepted as DLP-criteria. Familial hyperlipoproteinaemias were found in 19.7 % of the cases. Of these cases 5 % had hypo-alpha-lipoproteinaemia (see Table 7).

Table 7. Incidence of various types of hyperlipidaemias in men age between 20–59 years living in Moscow

Types of DLP	DLP criteria (mg/dl) CH	TG	HDL-CH	DLP all Observations	(%)	DLP with IHD (%)	Incidence of atherogenic DLP (%)
2a	>260	<200	>35	290	9.1	1.9	1:55
2b	>260	>200	>35	123	3.8	0.5	1:200
4	<260	>200	>35	61	1.9	0.56	1:166
Hypo-alpha	<260	<200	<35	156	4.9	0.8	1:125
All				630	19.7	2.76	1:36

According to strict criteria (Rose and Blackburn, 1968) IHD was found in 337 men out of a total 3201 (10.5 %). One-third of those with IHD symptoms had DLP, using the criteria shown above. At the same time 2.76 % of men with DLP had IHD (atherogenic DLPs). Our attention was drawn by the fact that almost half of IHD cases had been revealed in men aged between 20–49 years with DLPs.

Atherogenic DLPs in men aged 20–59 amount to:

1.9 % (1 : 55) in the case of 2b phenotype;
0.5 (1 : 200) in the case of 4 phenotype;
0.8 (1 : 125) in the case of hypo-alpha-lipoproteinaemia.

The structure of risk factors in DLP cases was studied by means of GLM with the following STEPWISE regression analysis. Types of DLPs were analysed. A specific combination of risk factors is essential with any DLP type (see Table 8).

Deaths because of stroke or heart attack, and their occurrence in anamnesis of parents, happened to be quite independent risk factors of DLP in men aged between 20–59 years living in Moscow. There are differences in the distribution of familial anamnesis data shown, depending upon the DLP type (see Table 9).

Types 2a and 2b hyperlipidemia posses high enough statistical significance for inclusion in the common linear model in the cases where there are indications

Table 8. Definition of connection between DLPs, normal lipid status and familial anamnesis data in men aged between 20–59 years living in Ashchabad and Moscow

Traits	Hyperlipoproteinaemia		Normal lipid status	
	F value	T criterial	F value	T criteria
Patient Data				
Age	134.35	0.0001	115.68	0.0001
Systolic blood pressure	13.06	0.0003	9.70	0.0019
Diastolic blood pressure	42.87	0.0001	25.92	0.0001
Body mass index	87.86	0.0001	164.58	0.0001
Cholesterol	1434.45	0.0001	199.16	0.0001
Triglycerides	102.12	0.0001	309.6	0.0001
HDL-Cholesterol	30.21	0.0001	180.09	0.0001
Mother's anamnesis				
Heart attack	0.24	0.7903	0.16	0.8517
Stroke	2.57	0.0617	2.40	0.0651
Cause of mother's death				
Heart attack	8.78	0.0093	4.98	0.0258
Stroke	4.63	0.0315	1.09	0.2975
Father's anamnesis				
Heart attack	3.26	0.0003	7.64	0.0005
Stroke	1.07	0.3420	0.53	0.5871
Cause of father's death				
Heart attack	12.20	0.0005	10.03	0.0013
Stroke	0.11	0.7369	0.00	0.9541
PR > F	0.0001	0.0001		
R square	39.32	37.3		

Table 9. Levels of T criteria in familial anamnesis data in various types of hyperlipidaemiae

Event	DLP	Normolipidaemia	2a	2b	4	Hypo-HDL-cholesterolemia
Stroke as cause of mother's death	0.03	–	0.04	–	–	–
Heart attack as cause of mother's death	0.009	0.002	0.005	0.05	–	–
Stroke as cause of father's death	–	0.05	–	–	–	–
Heart attack as cause of father's death	0.0003	0.0001	0.0009	–	–	–
Total	3	3	3	1	–	–

that the mother's and father's deaths were due to heart attack or stroke. At the same time, neither of the familial anamnesis traits in the cases of men with type 4 hyperlipidaemia or hypo-alpha-lipoproteinemia possessed any statistical significance for inclusion in the linear model.

Revealed IHD polymorphism combined with lipid metabolism disturbances (or without them) indicates the variability of alternate genetic factors which participate in pathogenesis. Observations vividly demonstrate the major role of lipid metabolism disturbances in IHD pathogenesis.

Genetic Epidemiological Features of Arterial Blood Pressure

The following populations were examined in a study of hereditary factors influencing the level of arterial blood pressure:

1. Male inhabitants, aged between 20–59 years from Moscow and Ashchabad city.
2. The population of the Nochur settlement in the Ashchabad area of Turkmenskaya USR.

We were concerned by the fact that distributions of arterial blood pressure levels did not differ significantly in all age groups in Moscow and Ashchabad. By contrast, in dwellers of the Nochur settlement, levels of systolic arterial blood pressure exceeded results from other populations by almost 10 mmHg, whereas diastolic blood pressure did not differ significantly.

It should be noted that comparative analysis of indexes of linear age regressions had its own features for every age group, both in men and in women living in Moscow and in Nochur. The growth of arterial blood pressure level is more rapid in the Nochur population than in the Moscow inhabitants, as can be seen in Fig. 1.

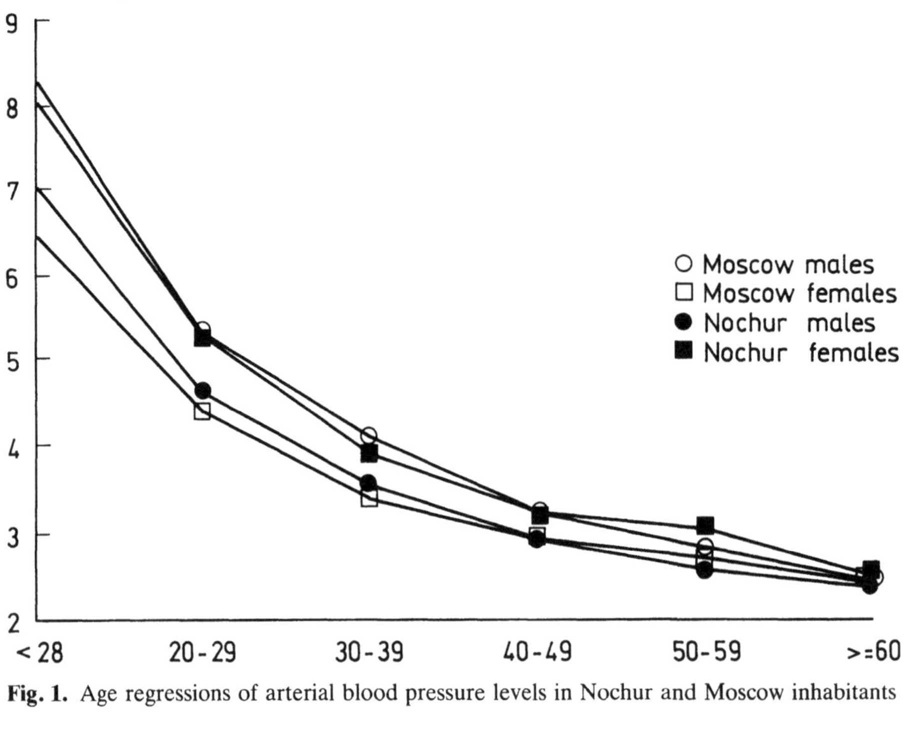

Fig. 1. Age regressions of arterial blood pressure levels in Nochur and Moscow inhabitants

Arterial blood pressure, age, BMI, and plasma lipid concentrations differ significantly in groups with HBP compared to those without it. Considerable similarity is found in the character of distribution of traits mentioned above in these populations (see Table 10).

Familial anamnesis data – heart attack and stroke as causes of death or mention of them among parents in family anamnesis – correlated statistically, correlated most highly with HBP in men who lived in Moscow or Ashchabad (see Table 11).

Table 10. Distribution (X ± m) of risk factors in men aged between 20–59 years living in Ashchabad and Moscow with and without HBP

Traits	Ashchabad			Moscow		
	HBP−	HBP+	Statistical significance	HBP−	HBP+	Statistical significance
	Y ± m	X ± m	($p<0.05$)a	X ± m	X ± m	($p<0.05$)a
Age	36 ± 0.2	43 ± 0.3	0.0001	38 ± 0.2	47 ± 0.5	0.0001
Systolic blood pressure	121 ± 0.3	150 ± 0.8	0.0001	124 ± 0.3	155 ± 1.3	0.0001
Diastolic blood pressure	79 ± 0.2	100 ± 1.1	0.0001	82 ± 0.2	101 ± 0.7	0.0001
Body mass index	24 ± 0.07	28 ± 0.2	0.0001	24 ± 0.1	27 ± 0.2	0.0001
Cholesterol	–	–	–	206 ± 0.8	221 ± 2.6	0.0001
Triglycerides	–	–	–	98 ± 1.2	126 ± 16.3	n. s.
HDL-Cholesterol	–	–	–	50 ± 0.2	51 ± 0.12	0.01

Symbols: a, T-Student criteria; n. s., not significant

Table 11. Distribution (%) of family anamnesis data in men aged between 20–59 years living in Ashchabad and Moscow with and without HBP

Traits	Ashchabad			Moscow		
	HBP−	HBP+	Statistical significance	HBP−	HBP+	Statistical significance
	X ± m	X ± m	($p<0.05$)a	X ± m	X ± m	($p<0.05$)a
Cause of mother's death						
Stroke	2.8 ± 0.4	7.9 ± 1.2	0.0001	4.7 ± 0.4	7.9 ± 0.9	0.01
Heart attack	3.0 ± 0.4	4.2 ± 0.9	n. s.	4.6 ± 0.4	6.8 ± 0.8	0.05
Cause of father's death						
Stroke	3.4 ± 0.4	4.6 ± 0.9	n. s.	4.5 ± 0.4	7.4 ± 0.9	n. s.
Heart attack	4.4 ± 0.5	5.9 ± 1.0	n. s.	4.7 ± 0.4	11.3 ± 1.1	0.01
Mother's anamnesis						
Stroke	4.6 ± 0.5	11.5 ± 1.4	0.001	8.5 ± 0.6	15.8 ± 1.3	0.05
Heart attack	16.5 ± 0.9	24.4 ± 2.0	0.001	27.8 ± 1.0	31.5 ± 1.7	n. s.
Father's anamnesis						
Stroke	5.4 ± 0.5	7.0 ± 1.2	n. s.	10.6 ± 0.7	13.8 ± 1.3	0.05
Heart attack	12.3 ± 0.8	15.2 ± 1.8	n. s.	24.4 ± 1.0	31.5 ± 1.0	0.001

Symbols: a, T-Student criteria; n. s., not significant

The determination of linkage between HBP manifestation and risk factors including familial anamnesis data was performed by means of the GLM procedure (see Table 12). Heart attack in mother's and father's anamnesis and stroke in mother's anamnesis are independent factors for inclusion in the general linear model to have HBP in males living in Ashchabad, whereas only heart attack in mother's anamnesis is the independent risk factor of HBP in males living in Moscow.

Table 12. Definition of connection of HBP manifestations with risk factors in men aged between 20–59 years living in Moscow and Ashchabad

Traits	Ashchabad		Moscow	
	F value	T criteria	F value	T criteria
Patient Data				
Age	253.71	0.0001	379.30	0.0001
Body mass index	131.47	0.0001	172.56	0.0001
Mother's anamnesis				
Heart attack	0.59	0.0035	2.47	0.0848
Stroke	7.62	0.0001	0.97	0.4061
Cause of mother's death				
Heart attack	0.35	0.5534	1.82	0.1774
Stroke	0.63	0.4292	1.37	0.2420
Father's anamnesis				
Heart attack	2.89	0.0338	2.60	0.0745
Stroke	0.34	0.8016	3.30	0.0370
Cause of father's death				
Heart attack	0.03	0.8713	0.20	0.6578
Stroke	0.29	0.5932	1.51	0.2198
PR < F	0.0001	0.0001		
R-square	39.32	37.36		

1. Equation of multiple regresion for men aged between 20–59 years living in Ashchabad:
   $y = 0.3327 + 0.009\ (\text{age}) + 0.0080\ (\text{BMI}) + 0.1100\ (\text{mother's death by stroke}) + 0.1288\ (\text{father's death by heart attack})$.
2. Equation of multiple regression for men aged between 20–59 years living in Moscow:
   $y = 0.4183 + 0.0064\ (\text{age}) + 0.0107\ (\text{BMI}) + 0.0692\ (\text{mother's death by heart attack})$.

Collections of traits picked by the analysis mentioned above differ in men living in Moscow and in Ashchabad. Thus, in Ashchabad men, HBP correlates significantly with the mother's death due to stroke and the father's death by heart attack, while in Muscovites, only the mother's death by heart attack correlates with HBP more or less significantly.

Variability of Arterial Blood Pressure in the Nochur Population

The incidence of HBP in the Nochur inhabitants of about 45 years of age is shown in Table 13. Of those examined, 34.9 % (32.7 % of the men, 37.0 % of the women) had elevated levels of arterial blood pressure amounting to 160/95 mmHg and over, according to examination data corresponding to the age of 45.

Acquired correlation coefficients were applied for the breakdown of phenotypic dispersion of SBP and DBP into genetic and environmental components. In the case of SBP, the contribution of genetic factors was 27.3 % ± 6.5 %, and in the case of DBP, 23.1 % ± 8.4 % (see Table 14). In the case of SBP, a considerable total environment component was revealed, 12.7 % ± 4.3 %.

The existence of a total environment component can possibly be explained by peculiarities in the mineral composition of the water. This water is used by Nochur inhabitants for drinking and everyday needs. Water is drawn from eight different springs. The conjecture that water is a factor is supported by the fact that, in groups with and without HBP microelement compositions of water differ considerably.

Calculations done by means of the GLM procedure show that age and spring number have a statistical significance for SBP, which is high enough for its

Table 13. Combined correlation coefficients in pairs of relatives in the Nochur population

Pairs of relatives	Number of pairs	Systolic		Diastolic	
		Coefficient	R = 0	Coefficient	R = 0
Father-son	346	0.0347	0.051	0.0502	0.051
Father-daughter	472	-0.0100	0.045	0.0145	0.045
Mother-daughter	556	0.1696	0.035	0.1156	0.037
Mother-son	406	0.1043	0.042	0.0877	0.048
Sister-sister	738	0.2635	0.083	0.1393	0.031
Brother-brother	299	0.2700	0.042	0.1055	0.051
Father-mother	249	0.0195	0.061	0.0685	0.059

Table 14. Breakdown of phenotypic dispersion of arterial blood pressure of the Nochur population into environmental and genetic components

Arterial blood pressure	Components of phenotypic dispersion							
	Genetic			Environmental				
				Environmental			Random	
	%	Stat. error	R = 0	%	Stat. error	R = 0	%	Stat. error
Systolic	27.3	6.5	0.01	12.7	4.3	0.03	59.8	7.8
Diastolic	23.1	8.4	0.01	2.4	5.5	0.4	77.5	10.0

inclusion in the HBP model for both men and women. However, for men, the spring number possessed a negative meaning in terms of the inclusion index, whereas in women it had a positive meaning. In the case of DBP, the spring number did not possess enough significance for inclusion in men, whereas for women it turned out to be significant. At the same time, age and spring number formed more than one-third of the residual dispersion which explained arterial blood pressure variability (see Table 15).

Table 15. Definition of connection of drinking water and phenotypic dispersion of blood pressure in Nochur inhabitants

Traits	T criteria			
	Males		Females	
	Systolic blood pressure	Diastolic blood pressure	Systolic blood pressure	Diastolic blood pressure
Spring	-0.0015	0.1061	0.0001	0.0001
Age	0.0001	0.0001	0.0001	0.0001
R-square (%)	51.7	34.3	44.8	36.4
PR > F	$p < 0.0001$	$p < 0.0001$		

Contribution of Genetic Factors to HBP Distribution in the Nochur Population

Segregation frequencies in siblings, separated according to HBP presence or absence in their parents, turned out to be as follows:

Parents:	Segregational frequency of HBP:
healthy * healthy	0.4658 ± 0.025
healthy * sick	0.4993 ± 0.022
sick * sick	0.5743 ± 0.032

Results of segregational frequencies in siblings reject the monogenic recessive model whether their parents suffered HBP or not. The monogenic dominant model is also rejected in families where both parents are healthy or suffer from HBP. A recessive model of HBP inheritance is not rejected in families where, according to segregational frequencies data only, one of the parents suffers HBP.

Conclusions

1. The incidence of IHD and HBP in populations is determined by the effect of genetic and environmental factors.

2. Familial anamnesis data on the existence of cardiovascular diseases in parents and siblings possess independent meaning as IHD and HBP risk factors. At the same time, familial anamnesis data possess as great a prognostic significance for IHD probability in men aged between 20–59 years as do more famour risk factors: elevated arterial blood pressure, body mass index, elevated concentrations of plasma lipids, etc.
3. Indication of cardiovascular disease among parents correlates significantly with various types of dislipoproteinaemias and high pressure in men aged between 20–59 years.
4. Interaction analysis of genetic and environmental factors in IHD manifestation performed by testing of a monolocus diallelic model in men aged between 20–59 years living in a Moscow district indicates that the model describes the examined sample well enough and in accordance with a quasidominant inheritance variant with incomplete penetrance of AA and Aa genotypes.
5. Acquired data testify that interindividual variability of cholesterol, triglyceride, and HDL-cholesterol concentrations is determined by a multitude of genetic and environmental factors.
6. In men aged between 20–59 years living in Moscow, dislipoproteinaemias of the 2a atherogenic type are 1 : 55; of the 2b type, 1 : 200; of the 4 type, 1 : 166; and hypo-alpha-lipoproteinaemia, 1 : 125. Revealed polymorphism of IHD manifestation combined with lipid metabolism disturbances (or without them) provides strong evidence supporting the participation of multivariate genetic factors in pathogenesis.
7. Among all the independent variables picked for examination – age, arterial blood pressure, body mass index, concentrations of lipids – only familial anamnesis data possessed statistically significant traits for inclusion in the regression model of dislipoproteinaemia probability.
8. Familial anamnesis data – heart attack and stroke as causes of death or mention of them in familial anamnesis of parents – were most highly statistically associated with increased levels of arterial blood pressure in men living in Moscow.
9. Determination of segregational frequencies among siblings from Nochur families (groups of siblings were formed based on HBP existence in their parents) suggests that HBP is inherited in accordance with the recessive autosomal model.
10. Results of the study dealing with IHD and HBP incidence provide strong evidence of the important role of genetic factors in IHD and HBP pathogenesis, a fact to be reckoned with in future studies of these widespread diseases.
11. Data from the study support the development of a new trend in genetic-epidemiological studies of cardiovascular diseases, focusing on investigation of the interaction of the features of genetics and environment in IHD and HBP pathogenesis in families and populations.

References

Berg K (1987) Genetics of coronary heart disease and its risk factors. In: Molecular approaches to human polygenic diseases. Ciba Foundation Symposium 130, Chichester, p 14–43
Chazov EI, Klimova M (1980) Dislipiproteidemii i ishemicheskaya bolezn serdtca. Medicina, 286 s.
Chasov EI, Repin VS, Orekhov AN et al. (1986). What has been learned studying human arteries. Atherosclerosis Res 14: p 6–7
Fredrickson D, Levy RJ (1972) Familial hyperlipoproteinaemia. In: Wyngaarden IB, Fredrickson D (eds) The metabolic basis of inherited diseases. McGraw Hill, New York, p 546–614
Koshechkin V. A. Geneticheskij analiz lipoproteidov v strukture nasledstvennoj predraspoloozennosti k ishemicheskoj bolexni serdtoa. Diss. doct. med. nauk, 1983. - M.
Koshechkin V. A. Perova N. V., Chepurnenko N. V., Trubnikov V. I., Sostojanie j perspectivi geneticheskih isolodovanij ishemicheskoj Bolezni serdtoa - Vestnik AMN SSSR, 1980, N 9, C. 2533.
Oganov R. G. Pervichnaya profilactika ishemicheskoj bolezni serdtca. M. Medicina. 1990, 140 s.
Rose GA, Blackburn H (1968) Cardiovascular survey methods. WHO Monography Series. No 56, p 190
WHO (1985) Primary prevention of essential hypertension. WHO Techn. Rep. Series No 686, - Geneva.
WHO (1986) Community prevention and control of cardiovascular diseases. WHO Technical Report Series. No 732, Geneva, p 62
Williams RR, Hunt SC, Barlow GK et al. (1988) Health family trees: a tool for finding and helping family members of coronary and cancer prone pedigrees in Texas and Utah. Am J Public Health. 78: p 1283–1286

The Role of Lp(a) in Atherosclerosis

R. Lawn

Summary

Lipoprotein(a) [Lp(a)] is a major, inherited risk factor for cardiac heart disease. Among individuals, Lp(a) levels vary nearly one-thousand fold, yet change very little during a person's lifetime. Lp(a) is distinguished from low density lipoprotein by the addition of the large glycoprotein apolipoprotein(a) [apo(a)]. The sequence of apo(a) is highly similar to plasminogen, suggesting some shared properties of the proteins, and pointing to its evolution via duplication of the plasminogen gene. Plasma Lp(a) concentration is largely inherited in a Mendelian codominant mode at the locus for the apo(a) structural gene. The gene is largely confined to higher primates, although there are now reports of Lp(a) in two sub-primate species. Although its similarity ot plasminogen may indicate a link between atherosclerosis and thrombosis, the precise physiological properties of apo(a) remain unknown.

Introduction

Atherosclerosis is the leading cause of death in the United States, yet the mechanisms of this disease are complex and poorly understood. Exhaustive epidemiologic, genetic and animal studies have uncovered a number of inherited and environmental risk factors for the disease. In the last few years, a novel lipoprotein species has been shown to account for a very large percentage of the remaining unspecified risks for atherosclerosis and myocardial infarction. This lipoprotein is known as lipoprotein(a), or Lp(a). Since its initial discovery and risk assessment by Kåre Berg and colleagues, it has been amply demonstrated that individuals in the top quartier of plasma Lp(a) levels have a 2- to 5-fold increased risk of heart disease and that elevated Lp(a) represents the most prevalent inherited risk factor for atherosclerosis (Berg 1963; Rhoads et al. 1986; Genet et al. 1989; Scanu and Fless 1990; K. Berg, this volume; R. Williams et al., this volume).

In contrast to most other risk factors, plasma levels of Lp(a) are insensitive to diet and drug manipulations (of all cholesterol lowering drugs, only high doses of niacin have been shown to lower Lp(a) levels). Understanding and ultimately influencing plasma levels of Lp(a) are of great public health interest. Lp(a)

K. Berg, V. Bulyzhenkov, Y. Christen, P. Corvol (Eds.)
Genetic Approaches to Coronary Heart Disease
and Hypertension

exerts its pathological effect at plasma levels of about 20–30 mg/dl, which is about one-fifth the level of low density lipoprotein (LDL) that is deemed significant. Yet Lp(a) closely resembles LDL in its content of cholesterol, phospholipid and apolipoprotein B-100, which is the ligand of the LDL receptor. The distinguishing feature of Lp(a) is the presence of an additional protein known as apolipoprotein(a) [apo(a)], bound to apolipoprotein B-100 by disulfide linkage (see Scanu and Fless 1990, for review and references therein).

Structure of Lp(a)

Figure 1 shows the structure of Lp(a), with its distinguishing moeity, apo(a), bound to apolipoprotein B-100 on the surface of the lipid-containing particle. To help elucidate the structure and activity of this unique protein, we undertook the cloning and DNA sequence analysis of apo(a) cDNA. With the collaboration of A. Scanu's laboratory at the University of Chicago, our groups at Genentech were able to obtain partial protein sequence of apo(a) which allowed the design of oligonucleotide probes and the identification of apo(a) clones from cDNA libraries constructed from human liver mRNA (Eaton et al. 1987; McLean et al. 1987).

Complete sequence analysis of a 14,000 base pair DNA copy of apo(a) mRNA had several surprises (McLean et al. 1987). The first was the occurrence of

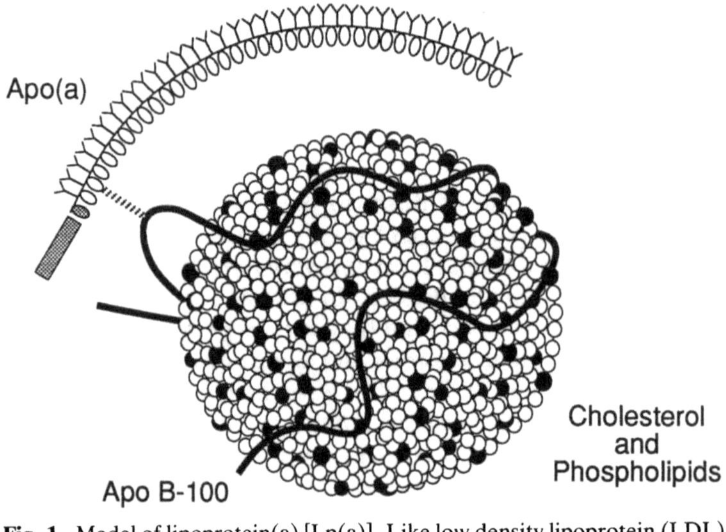

Fig. 1. Model of lipoprotein(a) [Lp(a)]. Like low density lipoprotein (LDL), Lp(a) consists of a core of cholesteryl esters and a surface layer of phospholipids and free cholesterol, plus one molecule apolipoprotein B-100. Lp(a) differs from LDL by the addition of the large glycoprotein apolipoprotein(a) (apo(a)) which is bound to B-100 by disulfide linkage. Apo(a) is related by sequence to plasminogen. It contains numerous (and variable among individuals) copies of a domain resembling the fourth kringle of plasminogen; in the case of apo(a), these domains are modified by sugar residues. At its carboxyl terminus, apo(a) has single copies of sequences which resemble the fifth kringle and the protease domain of plasminogen

numerous exact or nearly exact repeats of a 342 base sequence. Indeed, most of the mRNA was found to consist of 22 tandem exact repeats and 15 modified repeats. Although a number of proteins contain internally "duplicated" regions, the extent and fidelity of the repeated sequences in apo(a) were unprecedented. The second unexpected result was that the sequence of apo(a) closely resembles that of plasminogen. Plasminogen is a protease zymogen whose active form cleaves fibrin to dissolve blood clots. It is activated by tissue and urokinase plasminogen activators (tPA, UK) by cleavage at a specific arginine residue. Plasminogen consists of a preactivation peptide, five domains known as "kringles," and a catalytic portion that is related to the serine protease trypsin. The five kringles of plasminogen (which share only 50 % sequence identity with one another) are involved in binding to its fibrin substrate and to specific inhibitory proteins. It is the fourth kringle of plasminogen which most closely resembles the highly repeated domain of apo(a).

Figure 2 shows a pictorial comparison of human apo(a) and plasminogen mRNA and protein sequences. Apo(a) is a much larger protein than plasminogen, but all of its sequences have related counterparts in plasminogen. Both proteins have identical secretion signal peptides. Plasminogen contains an amino terminal preactivation region followed by three kringles which have no close counterpart in apo(a). The fourth and fifth kringle and protease domains have been conserved in apo(a). The originally described apo(a) cDNA contains 37 copies of a sequence which ranges from 75–85 % nucleotide and 61–75 % amino acid identity to plasminogen kringle four, and single domains even more closely related to kringle five and the protease regions of plasminogen. This unexpected similarity to plasminogen suggested activities in common with apo(a) that could provide clues to physiological functions of Lp(a) at low levels, as well as its pathological effects at elevated levels.

Inheritance of Lp(a)

Following their initial discovery Kåre Berg and coworkers conducted the first genetic studies that indicated that Lp(a) was an antigen inherited in an

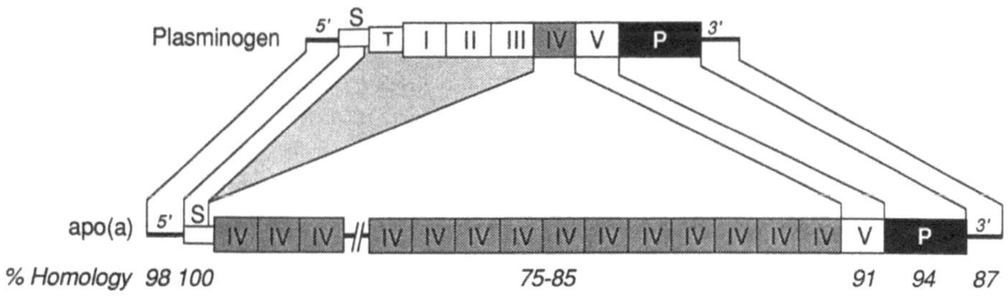

Fig. 2. Sequence structure comparison of plasminogen and apolipoprotein(a) [*apo(a)*]. The domains of plasminogen and apo(a) are compared. *S* indicates the signal peptide sequence; *T*, the tail or preactivation peptide; *I–V*, the kringle domains; and *P*, the protease. *Connecting lines* indicate regions of homology. The percentage of nucleotide sequence identity for each domain is indicated at the bottom. (From McLean et al. 1987.)

autosomal dominant fashion (Berg 1963; Berg and Mohr 1963). With the advent of quantitative assays for plasma Lp(a) concentration, and with the availability of polymorphic protein and DNA markers, linkage studies have put the inheritance of Lp(a) on a firm footing. There is a wide variation in the size of the apo(a) protein, and these isoforms serve as convenient polymorphic markers (Utermann et al. 1987). We proposed that out of register homologous recombination could lead to different numbers of identical kringle domain repeats in individual apo(a) genes, accounting for this size difference (McLean et al. 1987). Several groups have now confirmed this hypothesis (Hixson et al. 1989; Koschinsky et al. 1990; Gavish et al. 1989; Lindahl et al. 1990). It appears that from 15–40 kringle repeats are commonly found in apo(a) genes, leading to proteins that range from about 400,000 to 800,000 Daltons.

We found close linkage between RFLPs in the plasminogen gene, apo(a) protein isoforms, and plasma Lp(a) concentration (Drayna et al. 1988). Several other studies have also used plasminogen RFLPs or protein isoforms to show that plasminogen is closely linked to apo(a), and that apo(a) isoforms and Lp(a) plasma concentration are inherited in a Mandelian codominant mode (Weitkamp et al. 1988; Drayna et al. 1988; Utermann et al. 1988; Lindahl et al. 1989; Berg 1990). Although other gene products undoubtedly play a role in determining Lp(a) plasma levels, this represents an extreme example of heritability of a widely varying plasma component. Lp(a) levels vary one-thousand fold in the human population, yet it has been calculated that approximately half of this variability resides at the apo(a) gene locus (Boerwinkle et al. 1989). A second candidate locus for significant effect upon Lp(a) concentration is the LDL receptor gene. Biochemical reports have been split on the possible masking effects of apo(a) on the receptor binding activity of apolipoprotein B-100 (Maartmann-Moe and Berg 1981; Floren et al. 1981; Armstrong et al. 1985; Krempler et al. 1983; Hofmann et al. 1990). Higher mean Lp(a) levels were found in a series of British patients with the LDL receptor defect, familial hypercholesterolemia (FH), than in healthy Tyrolian control subjects (Utermann et al. 1989). These investigators also found that, among a group of British FH patients, those with coronary artery disease had higher mean Lp(a) levels than those free from disease (Seed et al. 1990). Taken together, these studies indicate the important additive risk of FH and Lp(a), but this might also increase the plausibility of an individual's likelihood of appearing as a patient in a lipid clinic. In apparent contrast, Scanu and coworkers reported that members of a rhesus monkey pedigree with FH exhibited no correlation between LDL receptor function and plasma Lp(a) levels (Neven et al. 1990) and Soutar and colleagues (1991) recently made a similar observation in a human pedigree. Hence, the importance of the LDL receptor in determining plasma Lp(a) concentrations remains to be firmly established. Undoubtedly, factors other than the apo(a) gene itself play a role in determining plasma Lp(a) levels. But in contrast to most plasma macromolecular complexes, Lp(a) exhibits a very wide range in concentration in the general population, and much of this variation is determined by genetic, rather than environmental, factors.

Evolution of apo(a)

Apo(a) appears to be a recently evolved member of the protein superfamily of plasma proteases, which includes plasminogen and other members of the blood coagulation and fibrinolytic systems. These proteins contain protease domains which resemble trypsin, and non-catalytic domains that are involved in localization, activation and interaction with cofactors and inhibitors. Many new genes and proteins appear to arise by gene duplication, followed by sequence divergence to provide novel function, and by duplication or interchange of parts of genes to add pre-existing or modified domains to emerging proteins. Such scenarios have been proposed for the gene family that includes plasminogen and apo(a) (Patthy 1985; Castellino and Beals 1987). By virtue of its extreme homology to plasminogen, it is possible that the apo(a) gene arose from a common precursor rather recently. Indeed, based upon the nucleotide substitution rate calculated for the untranslated regions of primate globin genes (Maeda et al. 1983), the apo(a) gene may have arisen as recently as 40 million years ago, at about the time of separation of world and new world monkeys. Consistent with this hypothesis, Lp(a) has been reported in old world monkeys and apes, but is not found in new world monkeys or prosimians. It is reasonable to propose that the duplication of the plasminogen gene provided the genetic material to form the apo(a) gene. Sequences encoding the preactivation peptide and kringles one, two and three of plasminogen were eliminated. [And in the case of rhesus monkey Lp(a), the kringle five domain was eliminated as well (Tomlinson et al. 1989).] Sequence divergence has occurred, as well as the multiplication of the kringle four-like sequences. In addition, a sequence correction mechanism know as the gene conversion probably produced the nearly 100% sequence identity of the 5′ regions of the genes, as shown in Figure 2. It remains to be determined which of these sequence features is crucial for an adaptive function of apo(a). Comparison of apo(a) sequences from several species and in vitro mutagenesis of recombinant apo(a) should help elucidate this question.

The straightforward hypothesis that the apo(a) gene arose during primate evolution may need re-evaluation, because, despite the evidence from sequence comparison and direct testing of plasma for the confinement of apo(a) to higher primates, the lipoprotein has now been reported to exist in hedgehogs and guinea pigs (Laplaud et al. 1988; Rath and Pauling 1990). Cold the apo(a) gene really predate primate evolution? One would then have to postulate its loss from nearly all mammals except higher primates, and a few exceptional sub-primate species. In addition, extreme selection pressure and gene conversion would have to be evoked to account for the extreme similarity of plasminogen and apo(a) sequences that we see in today's humans. An alternate explanation seems reasonable. It would be quite unlikely for evolution to produce the same gene twice from scratch. But perhaps all one needs to re-create functional apo(a) is a plasminogen gene (which exists in all mammals) and the chance to duplicate it and allow some modifications which give the new protein the ability to bind to apolipoprotein B-100 and thus become part of an LDL-like particle which has some useful function for the organism.

Function of Lp(a)

Although the link between elevated plasma Lp(a) levels and the risk for atherosclerosis is well established, its functions remain obscure. The unexpected similarity of apo(a) to plasminogen could provide a link between atherosclerosis and thrombosis. It suggests common activities that could provide clues to physiological functions of Lp(a) at low levels, as well as its pathological effects at elevated levels. For example, if apo(a) could mimic plasminogen in its ability to bind fibrin or other macromolecules at the sites of vessel damage, the delivery of cholesterol associated with the particle could lead to the development of atherosclerosis. Indeed, it has been speculated that the reson for the evolution of apo(a) could be to deliver cholesterol for cell membrane biosynthesis as part of the wound healing process (Brown and Goldstein 1987). In vitro studies of the activites of Lp(a) are underway in a number of laboratories but have yet to lead to many conclusions that are generally accepted in the field.

Figure 3 depicts some of the ways in which Lp(a) might become associated with a developing plaque. Plasminogen binds its substrate fibrin chiefly through a lysine binding site in its first kringle. Kringle four binds fibrin more weakly, but the many copies of a homologous kringle four-like domain in apo(a) might be expected to lead to tight binding to fibrin. However, there are key amino acid substitutions between the proteins, and in several in vitro settings, avid binding to fibrin is not detected. However, partial plasmin degradation of fibrinogen leads to enhanced apo(a) binding, and partially degraded fibrin clots might be a reasonable physiologic substrate (Harpel et al. 1989; Loscalzo et al. 1990). Lp(a) has been localized to atherosclerotic plaques, and binding to residual thrombi, as well as extracellular matrix components such as fibronectin and glycosamino-

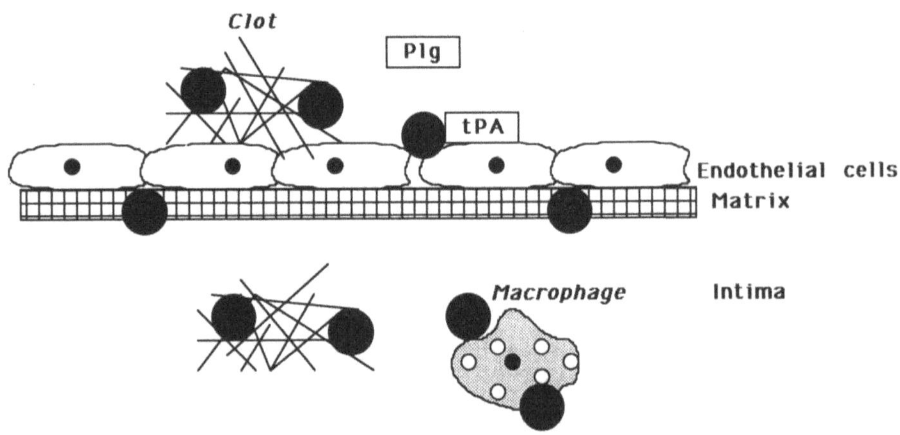

Fig. 3. Possible routes of lipoprotein(a) [Lp(a)] into the arterial wall. This simplified cartoon depicts the surface endothelial layer of the vessel, and subendothelial cell matrix and smooth muscle layers. Lp(a) (*filled circles*) can bind to matrix components or to residual fibrin clots, which can become incorporated into the growing plaque. It can compete with binding of plasminogen (*Plg*) to endothelial cells, inhibiting the effect of tissue plasminogen activator (*tPA*) and the subsequent lysis of fibrin. Lp(a) also binds to macrophages, which may contribute to the formation of foam cells

glycans, could be accountable (Rath et al. 1989; Cushing et al. 1989; Salonen et al. 1989; Bihara-Varga et al. 1988).

Lp(a) might exert an atherogenic effect by competing with plasminogen for binding to endothelial cells, thereby inhibiting the surface localization of plasminogen and tissue plasminogen activator and impeding fibrinolysis (Hajjar et al. 1989; Miles et al. 1989). This leads to the hypothetical scenario whereby elevated Lp(a) causes prolonged retention of residual thrombi at the vessel surface. Microthrombi can become incorporated in atherosclerotic plaque, perhaps due to growth factors released by platelets or other cells associated with thrombi. Recently, we have shown that macrophages may also play a role in the association of Lp(a) and the diseased vessel. Lp(a) binds to macrophages through a specific, high affinity receptor which appears to be distinct from the scavenger, or acetyl-LDL receptor (Zioncheck et al. 1991). Once bound to the surface of macrophages, and sequestered from circulating anti-oxidants, Lp(a) could become oxidized and ingested, converting macrophages into lipid-laden foam cells. Foam cells have been proposed to be atherogenic by a number of means, including chemotaxis, release of mitogenic catabolites and cytotoxic oxygen radicals, and by the induction of tissue factor (see Steinberg et al. 1989, for review).

In the 25 years since its discovery by Berg and coworkers, Lp(a) has been shown to be one of the leading risk factors for coronary heart disease, perhaps the most prevalent inherited risk. We have a good understanding of its structure and its genetics, but much remains to be learned about its role in normal body function, and its means of pathogenicity at elevated levels.

References

Armstrong VW, Walli AK, Seidel D (1985) Isolation, characterization, and uptake in human fibroblasts of an apo(a)-free lipoprotein obtained on reduction of lipoprotein(a). J Lipid Res 26: 1314–1323

Berg K (1963) A new serum type system in man: the Lp system. Acta Pathol Microbiol Scand 59: 369–382

Berg K (1990) Genetics of atherogenic Lp(a) lipoprotein. In: Berg K, Retterstol N, Refsum S (eds) From phenotype to gene in common disorders. Munksgaard, Copenhagen, pp 138–162

Berg K, Mohr J (1963) Genetics of the Lp system. Acta Genet 13: 349–360

Bihara-Varga M, Gruber E, Rotheneder M, Zechner R, Kostner G (1988) Interaction of Lp(a) and low density lipoprotein with glycosaminoglycans from human aorta. Arteriosclerosis 8: 851–857

Boerwinkle E, Menzel HJ, Kraft HG, Utermann G (1989) Genetics of the quantitative Lp(a) lipoprotein trait. III. Contributions of Lp(a) glycoprotein phenotypes to normal lipid variation. Hum Genet 82: 73–78

Brown MS, Goldstein JL (1989) Teaching old dogmas new tricks. Nature 232 (300): 113–114

Castellino FJ, Beals JM (1989) The genetic relationship between kringle domains of human plasminogen, prothrombin, tissue plasminogen activator, urokinase and coagulation factor XII. J Mol Evol 26: 358–369

Cushing GL, Gaubatz JW, Nava ML, Burdick BJ, Bocan TMA, Guyton JR, Weilbaecher D, DeBakey ME, Lawrie GM, Morrisett JD (1989) Quantitation and localization of apolipo-

protein(a) and B in coronary artery bypass vein grafts resected at reoperation. Arteriosclerosis 9: 593–603
Drayna DT, Hegele RA, Hass PE, Emi M, Wu LL, Eaton DL, Lawn RM, Williams RR, White RL, Lalouel JM (1988) Genetic linkage between lipoprotein(a) phenotype and a DNA polymorphism in the plasminogen gene. Genomics 3: 230–236
Eaton D, Fless G, Kohr W, McLean J, Xu Q, Miller C, Lawn R, Scanu A (1987) Partial amino acid sequence of apolipoprotein(a) shows that it is homologous to plasminogen. Proc Natl Acad Sci USA 84: 3224–3228
Floren CH, Albers JJ, Bierman EL (1981) Uptake of Lp(a) lipoprotein by cultured fibroblasts. Biochem Biophys Res Commun 102: 636–639
Gavish D, Azrolan N, Breslow J (1989) Plasma Lp(a) concentrations is inversely correlated with the ratio of kringle IV/kringle V encoding domains in the apo(a) gene. J Clin Invest 84: 2021–2027
Genest J, Martin-Munley S, McNamara J, Salem D, Schaefer E (1989) Frequency of genetic dyslipidemias in patients with premature coronary artery disease. Circulation [Suppl II]: II-80
Hajjar KA, Gavish D, Breslow JL, Nachman RL (1989) Lipoprotein(a) modulation of endothelial cell surface fibrinolysis and its potential role in atherosclerosis. Nature 339: 303–305
Harpel PC, Gordon BR, Parker TS (1989) Plasminogen catalyzes binding of lipoprotein(a) to immobilized fibrinogen and fibrin. Proc Natl Acad Sci USA 86: 3847–3851
Hixson JE, Britten ML, Manis GS, Rainwater DL (1989) Apolipoprotein(a) glycoprotein isoforms result from size differences in apo(a) mRNA. J Biol Chem 264: 6013–6016
Hofmann SL, Eaton DL, Brown MS, McConathy WJ, Goldstein JL, Hammer RE (1990) Overexpression of human LDL receptors leads to accelerated catabolism of Lp(a) lipoprotein in transgenic mice. J Clin Invest 85: 1542–1547
Koschinsky ML, Beisiegel U, Henne-Bruns D, Eaton DL, Lawn RM (1990) Apolipoprotein(a) size heterogeneity is related to variable number of repeat sequences in its mRNA. Biochemistry 29: 640–644
Krempler F, Kostner GM, Roscher A, Haslauer F, Bolzano K, Sandhofer F (1983) Studies on the role of specific cell surface receptors in the removal of lipoprotein(a) in man. J Clin Invest 71: 1431–1441
Laplaud PM, Beaubatie L, Rall SJ, Luc G, Saboureau M (1988) Lipoprotein(a) is the major apoB-containing lipoprotein in the plasma of a hibernator, the hedgehog. J Lipid Res 29: 1157–1170
Lindahl G, Gersdorf E, Menzel HJ, Duba C, Cleve H, Humphries S, Utermann G (1989) The gene for the Lp(a)-specific glycoprotein is closely linked to the gene for plasminogen on chromosome 6. Hum Genet 81: 149–152
Lindahl G, Gersdorf E, Menzel H, Seed M, Humphries S, Utermann G (1990) Variation in the size of human apolipoprotein(a) is due to a hypervariable region in the gene. Hum Genet 84: 563–567
Loscalzo J, Weinfeld M, Fless GM, Scanu AM (1990) Lipoprotein(a), fibrin binding, and plasminogen actication. Arteriosclerosis 10: 240–245
Maartmann-Moe K, Berg K (1981) Lp(a) lipoprotein enters cultured fibroblasts independently of the plasma membrane low density lipoprotein receptor. Clin Genet 20: 352–362
Maeda N, Bliska JB, Smithies O (1983) Recombination and balanced chromosome polymorphism suggested by DNA sequences 5′ to the human delta-globin gene. Proc Natl Acad Sci USA 80: 5012–5016
McLean J, Tomlinson J, Kuang W-J, Eaton D, Chen E, Fless G, Scanua A, Lawn R (1987) cDNA sequence of human apolipoprotein(a) is homologous to plasminogen. Nture 330: 132–137
Miles LA, Fless GM, Levin EG, Scanu AM, Plow EF (1989) A potential basis for the thrombotic risks associated with lipoprotein(a). Nature 339: 301–303
Neven L, Khalil A, Pfaffinger D, Fless GM, Jackson E, Scanu AM (1990) Rhesus monkey model of familial hypercholesterolemia: relation between plasma Lp(a) levels, apo(a) isoforms, and LDL-receptor function. J Lipid Res 31: 633–640

Patthy L (1985) Evolution of the proteases of blood coagulation and fibrinolysis by assembly from modules. Cell 41: 657–663
Rath M, Niendorf A, Reblin T, Dietel M, Krebberr HJ, Beisiegel U (1989) Detection and quantitation of lipoprotein(a) in the arterial wall of 107 coronary bypass patients. Arteriosclerosis 9: 579–592
Rath M, Pauling L (1990) Hypothesis: Lipoprotein(a) is a surrogate for ascorbate. Proc Natl Acad Sci USA 87: 6204–6207
Rhoads G, Dahlen G, Berk K, Morton N, Dannenberg A (1986) Lp(a) lipoprotein as a risk factor for myocardial infarction. JAMA 256: 2540–2544
Salonen E, Jauhianen L, Zardi L, Vaheri A, Enholm C (1989) Lipoprotein(a) binds to fibronectin. EMBO J 8: 4035–4040
Scanu A, Fless A (1990) Lipoprotein(a). J Clin Invest 85: 1709–1715
Seed M, Hoppichler F, Reaveley D, McCarthy S, Thompson GR, Boerwinkle E, Utermann G (1990) Relation of serum lipoprotein(a) concentration and apolipoprotein(a) phenotype to coronary heart disease in patients with familial hypercholesterolemia. N Engl J Med 322: 1494–1499
Soutar AK, McCarthy SN, Seed M, Knight BL (1991) Relationship between apolipoprotein(a) phenotype, lipoprotein(a) concentration in plasma and low density lipoprotein (LDL) receptor function in a large kindred with familial hypercholesterolaemia due to the pro664-leu mutation in the LDL-receptor gene. J Clin Invest 88: 483–492
Steinberg D, Parthasarathy S, Carew TE, Khoo JC, Witztum JL (1989) Beyond cholesterol: modifications of low-density lipoprotein that increase its atherogenicity. Engl J Med 320: 915–9245
Tomlinson J, McLean J, Lawn R (1989) Rhesus monkey apolipoprotein(a). J Biol Chem 264: 5959–5965
Utermann G, Menzel HJ, Kraft HG, Duba HC, Kemmier HG, Seitz C (1987) Lp(a) glycoprotein phenotypes. J Clin Invest 80: 458–465
Utermann G, Duba C, Menzel H (1988) Genetics of the quantitative Lp(a) lipoprotein trait. II. Inheritance of Lp(a) glycoprotein phenotypes. Hum Genet 78: 47–50
Utermann G, Hoppichler F, Dieplinger H, Seed M, Thompson G, Boerwinkle E (1989) Detects in the low density lipoprotein receptor gene affect lipoprotein(a) levels: multiplicative interaction of two gene loci associated with premature atherosclerosis. Proc Natl Acad Sci USA 86: 4171–4174
Weitkamp LR, Guttormsen SA, Schultz JS (1988) Linkage between the loci for the Lp(a) lipoprotein (LP) and plasminogen (PLG). Hum Genet 79: 80–82
Zioncheck TF, Powell LM, Rice GC, Eaton DL, Lawn RM (1991) Interaction of recombinant apolipoprotein(a) and lipoprotein(a) with macrophages. J Clin Invest 87: 767–771

Liver-Directed Gene Therapy In the Treatment of Familial Hypercholesterolemia

J. M. Wilson and *M. Grossman*

Gene replacement therapy is being considered for the treatment of numerous inherited and acquired diseases (Anderson 1984; Ledley 1987a, 1987b). Bone marrow directed gene therapy has been the paradigm for investigation in this area because the appropriate target cell for gene transfer, the hematopoietic stem cell, is easily accessed, cultured, and transplanted (Lehn 1987). Many diseases, however, will require genetic correction in somatic tissue other than bone marrow, such as liver, the central nervous system, airway respiratory epithelial cells, smooth muscle, and skeletal muscle just to name a few. We have been developing gene replacement therapies based on the transfer of normal genetic material into hepatic tissues.

One approach that we have developed for the genetic treatment of liver specific inherited diseases is similar in concept to the well-described bone marrow gene therapies. This method, referred to as ex vivo gene therapy, involves isolating hepatocytes from an animal, transferring functional genetic material into the hepatocyte in vitro, and transplanting the genetically modified hepatocytes back into the same animal. Several significant technical problems have inhibited progress in this area. An extremely efficient gene transfer system is required because hepatocytes have limited proliferative potential and can be very difficult to maintain in culture. We and others have recently used recombinant retroviruses to stably transduce a variety of genes into primary cultures of hepatocytes (Wolff et al. 1987; Ledley et al. 1987; Wilson et al. 1988a, 1988b; Peng et al. 1988). Our method differs from the others in that efficient infection is obtained in the absence of selection, thereby, minimizing the time hepatocytes must be maintained in culture prior to transplantation (Wilson et al. 1988a, 1988b).

The crucial step in ex vivo gene therapy, hepatocyte transplantation, has also been difficult to perform. Earlier studies attempted to transplant limited numbers of syngeneic hepatocytes into ectopic sites such as spleen, peritoneal cavity, or renal subcapsule (Matas et al. 1976; Mito et al. 1979; Jirtle et al. 1980; Makowka et al. 1981). Cells with hepatocyte-like morphology were detected at the transplantation sites at various times following the initial transplant. However, the functional or systemic consequences of these cellular grafts has never been studied. One approach to this problem is to transplant normal hepatocytes into animals that have a genetic deficiency of an hepatocyte-specific enzyme, receptor, or secreted protein. Improvement in the biochemical or

K. Berg, V. Bulyzhenkov, Y. Christen, P. Corvol (Eds.)
Genetic Approaches to Coronary Heart Disease
and Hypertension

metabolic consequences of the deficiency state in transplant recipients is a direct measure of the function of the engrafted hepatocyte.

An alternative and possibly more feasible approach to liver-directed gene therapy is to deliver the gene of interest into the hepatocyte in vivo. This approach has the potential to be more effective and less morbid than the ex vivo approach described above. Previous attempts at targeting the delivery of conventional therapeutic agents to the liver, such as antibiotics and chemotherapeutic agents, were based on specific interactions with the asialoglycoprotein receptor (Wu 1988). This transmembrane protein is specifically expressed in hepatocytes at very high levels in vivo, approximately 500,000 receptors/cell (Ashwell and Harford 1982). The asialoglycoprotein receptor binds and degrades plasma glycoproteins which have lost a significant number of terminal sialic acid residues.

Wu and Wu (1987, 1988) utilized the asialoglycoprotein receptor to develop a soluble method for transferring functional genetic material into hepatocytes in vivo. A high affinity ligand for the asialoglycoprotein receptor, asialoorosomucoid (ASOR), was covalently attached to polylysine and the protein conjugate was coupled noncovalently with a DNA expression vector. When injected into the venous circulation of rats, the protein/DNA adduct was specifically internalized by hepatocyctes via the asialoglycoprotein receptor and the reporter gene of the vector was transiently expressed at low levels.

An inherited deficiency of the receptor for LDL has been used as a model for developing liver-directed gene therapies. In humans, this genetic disease is called familial hypercholesterolemia (FH) and is inherited in an autosomal dominant manner (Goldstein and Brown 1983). Patients with one abnormal allele have moderate elevations in plasma LDL and suffer premature coronary artery disease; approximately 5 % of all patients under 45 who have had a myocardial infarction carry this trait. Patients with two abnormal LDL receptor alleles have severe hypercholesterolemia and life-threatening coronary artery disease in childhood. Heterozygotes can be managed clinically using conventional medical therapies. No effective therapy exists for homozygous-deficient patients other than orthotopic liver transplantation. It is these patients who may be considered for early trials of human gene therapy.

The liver is a primary organ responsible for regulation of cholesterol balance, and the receptor for LDL plays an important role in this regulation (Turley and Dietschy 1988). VLDL is a triglyceride-rich lipoprotein that is secreted by the liver and subsequently converted to IDL through the action of lipoprotein lipase in peripheral tissues. IDL has a high affinity for hepatic LDL receptors and is either degraded by those receptors or converted to a lipoprotein with decreased affinity for the LDL receptor (i.e., LDL). LDL is the primary carrier of cholesterol in the plasma and has been implicated in the development of atherosclerotic disease. Hepatic LDL receptors contribute to greater than 90 % of LDL degradation. Another important function of the liver in terms of cholesterol metabolism is that it is the only organ capable of providing a metabolic sink for cholesterol excretion. This is accomplished through the biosynthesis of bile acids from free cholesterol and the secretion of cholesterol-rich bile.

The mechanisms responsible for hypercholesterolemia in patients with FH have been well described (Goldstein and Brown 1983; Turley and Dietschy 1988). A deficiency of hepatic LDL receptors leads to an impairment in the catabolism of LDL. This receptor deficiency state also causes decreased degradation of the precursor of LDL (i.e., IDL) resulting in an overproduction of LDL. Based on our understanding of hepatic cholesterol metabolism and the pathophysiology of FH, we believe that the hepatocyte is the appropriate target for gene replacement therapy in this disorder. Previous attempts to treat patients with the homozygous deficient form of FH by liver transplantation has provided compelling evidence in support of this hypothesis. Transplantation of liver with normal LDL receptor function esentially normalizes serum LDL in these patients (Bilheimer et al. 1984).

In 1973, a strain of rabbits, called Watanabe Heritable Hyperlipidemic (WHHL) rabbits, that demonstrated clinical and metabolic abnormalities similar to those associated with FH in humans was identified (Watanabe 1980). The WHHL rabbit was subsequently shown to be deficient in functional LDL receptor activity due to a 12 bp inframe deletion of a portion of the LDL receptor structural gene (Yamamoto et al. 1988). The WHHL rabbit has been instrumental in delineating key steps in the regulation of cholesterol metabolism (Goldstein et al. 1983) and has been instrumental in the testing of our gene therapy strategies.

We have used the WHHL rabbit as a model for developing liver-directed gene replacement therapies that are effective and relatively nonmorbid. These studies have provided us with a basis for designing human gene therapies for FH. Most patients with moderate hypercholesterolemia have defects unrelated to the LDL receptor gene and can be managed with dietary restriction and/or intensive medical therapy. Over-expression of hepatic LDL receptors by gene therapy could potentially provide a cure in these patients, thereby obviating the need for ongoing pharmacological therapy.

The feasibility of ex vivo gene therapy has been demonstrated in the WHHL rabbit. Hepatocytes from a WHHL rabbit were isolated by collagenase perfusion, plated as primary cultures, and exposed to recombinant retroviruses capable of efficiently transferring a functional human LDL receptor gene. Genetically modified cells were harvested and infused into the portal vein of WHHL recipients, who were analyzed for metabolic consequences of human LDL receptor expression. Each animal exhibited a statistically significant decrease in total serum cholesterol 2–6 days after transplantation, with an eventual return to baseline levels. Proviral DNA sequences and viral-directed transcripts were detected in liver tissue 10 min and 24 h after transplantation. In situ hybridization demonstrated provirus expression in a small population of hepatocytes distributed in periportal sections of the liver (Wilson et al. 1990).

These encouraging results in the WHHL rabbit support the feasibility of somatic gene therapy in the treatment of FH. Several additional milestones should be achieved before human trials are initiated. First, the ex vivo protocol should be modified to allow prolonged, if not permanent, correction of hyperlipidemia. In addition, methods for isolating and efficiently transducing human hepatocytes must be developed. Much needs to be accomplished before gene therapy of FH is a reality. However, the path we need to follow to reach this goal is clearly charted.

References

1. Anderson WF (1984) Prospects for human gene therapy. Science 226: 401–409
2. Ashwell G, Harford J (1982) Carbohydrate-specific receptors of the liver. Ann Rev Biochem 51: 531–554
3. Bilheimer DW, Goldstein JL, Grundy SM, Starzl TE, Brown MS (1984) Liver transplantation to provide low density lipoprotein receptors and lower plasma cholesterol in a child with homozygous familial hypercholesterolemia. New Engl J Med 311: 1658–1664
4. Goldstein JL, Brown MS (1983) In: Stanbury JB, Wyngaarden JB, Fredrickson DS, Goldstein JL, Brown MS (eds) Metabolic basis of inherited diseases. McGraw-Hill, New York, pp 672–712
5. Goldstein JL, Kita T, Brown MS (1983) Defective lipoprotein receptors and atherosclerosis. New Engl J Med 309: 288–296
6. Jirtle RL, Biles C, Michalopoulos G (1980) Morphologic and histochemical analysis of hepatocytes transplanted into syngeneic hosts. Am J Pathology 101: 115–122
7. Ledley FD (1987a) Somatic gene therapy for human disease: background and prospects. Part I. J Pediatrics 110: 1–8
8. Ledley FD (1987b) Somatic gene therapy for human disease: background and prospects. Part II. J Pediatrics 110: 167–174
9. Ledley FD, Darlington GJ, Hahn T, Woo SLC (1987) Retroviral gene transfer into primary hepatocytes: implications for genetic therapy of liver-specific functions. Proc Natl Acad Sci USA 84: 5335–5339
10. Lehn PM (1987) Gene therapy using bone marrow transplantation. Bone Marrow Transplantation 1: 243–258
11. Makowka L, Rotstein LE, Falk RE, Falk JA, Zuk R, Langer B, Blendis LM, Phillips MJ (1981) Allogeneic and xenogeneic hepatocyte transplantation. Transplantation Proceedings XIII: 855–859
12. Matas AJ, Sutherland DER, Steffes MW, Mauer SM, Lowe A, Simmons RL, Najarian JS (1976) Hepatocellular transplantation for metabolic deficiencies: decrease of plasma bilirubin in Gunn rats. Science 192: 892–894
13. Mito M, Ebata H, Kusano M, Onishi T, Saito T, Sakamoto S (1979) Morphology and function of isolated hepatocytes transplanted into rat spleen. Transplantation 28: 499–505
14. Peng H, Armentano D, MacKenzie-Graham L, Shen R-F, Darlington G, Ledley FD, Woo SLC (1988) Retroviral-mediated gene transfer and expression of human phenyl-alanine hydroxylase in primary mouse hepatocyles. Proc Natl Acad Sci USA 85: 8146–8150
15. Turley SD, Dietschy (1986) The metabolism and excretion of cholesterol by the liver. In: Arias IM, Jakoby WB, Popper H, Schacter D, Shafritz DA (eds) The liver: biology and pathobiology, 2nd edn. Raven, New York, pp 617–641
16. Watanabe Y (1980) Serial inbreeding of rabbits with hereditary hyperlipidemia (WHHL rabbit). Atherosclerosis 36: 261–268
17. Wilson JM, Jefferson DM, Chowdhury JR, Novikoff PM, Johnston DE, Mulligan RC (1988) Retrovirus-mediated transduction of adult hepatocytes. Proc Natl Acad Sci USA 85: 3014–3018
18. Wilson JM, Johnston DE, Jefferson DM, Mulligan RC (1988) Correction of the genetic defect in hepatocytes from the Watanabe heritable hyperlipidemic rabbit. Proc Natl Acad Sci USA 85: 4421–4425
19. Wilson JM, Chowdhury NR, Grossman M, Wajsman R, Epstein A, Mulligan RC, Chowdhury JR (1990) Temporary amelioration of hyperlipidemia in low density lipoprotein receptor-deficient rabbits transplanted with genetically modified hepatocytes. Proc Natl Acad Sci USA 87: 8437–8441

20. Wolf JA, Yee J-K, Skelly HF, Moores JC, Respess JG, Friedmann T, Leffert H (1987) Expression of retrovirally transduced genes in primary cultures of adult rat hepatocytes. Proc Natl Acad Sci USA 84: 3344–3348
21. Wu GY (1988) Targeting in diagnosis and therapy. In: Arias IM, Jakoby WB, Popper H, Schacter D, Shafritz DA (eds) The liver: biology and pathobiology, 2nd edn. Raven, New York, pp 1303–1313
22. Wu GY, Wu CH (1987) Receptor-mediated in vitro gene transformation by a soluble DNA carrier system. J Biol Chem 262: 4429–2243
23. Wu GY, Wu CH (1988) Receptor-mediated gene delivery and expression in vivo. J Biol Chem 263: 14621–14624
24. Yamamoto T, Bishop RW, Brown MS, Goldstein JL, Russell DW (1986) Deletion in cysteine-rich region of LDL receptor impedes transport to cell surface in WHHL rabbit. Science 232: 1230–1237

Subject Index

adopted children 2
AIDS 16
aldosterone 34, 42
allelic effects on biological markers of hypertention 31–34
ANF 27
angiotensin II 34, 74
angiotensin-I-converting enzyme 33, 34
angiotensin converting enzyme inhibitor 65, 76
angiotensinogen 33, 45, 46, 74
animal model of hypertention 6, 8, 60–72, 74–79, 81–95
anthropometric measurement 10
antiporter gene 6, 32
antiporter protein 5, 14
apolipoprotein 10, 99–102, 110, 111, 144
apolipoprotein a 110, 144, 145, 147
apolipoprotein A–I deficiency 111, 122–124
apolipoprotein B 12, 13, 98, 100, 101, 111, 118, 144, 147
apoliproprotein E 100, 104, 105, 110, 111, 114–117, 124
arteriolipidosis-prone rat (ALR) 81–83, 90
atherosclerosis 16, 81, 90, 91, 98, 102, 103, 114, 115, 122, 128, 143–149

Brown-Norway normotensive rat (BN) 38, 43, 45, 47–53

Ca^{++} homeostasis and hypertention 22, 25
candidate gene and hypertention 27–31, 41–48, 57, 98
cell membrane alteration in hypertention 20–26
cholesteryl ester transfer protein 101
chromosome 1 45
chromosome 5 (rat) 83
chromosome 13 (rat) 45
chromosome 19 16
chromosome X 56
chromosome Y 56
Church of Jesus Christ of Latter Day Saints (Mormons) 17
coronary heart disease 8–18, 98–126
c-src 23–25
cytoskeleton 23

diabetes mellitus 35
dietary factors 93, 94, 98, 103
dietary electrolyte intake 14
dislipoproteinaemias 134–136, 141

early coronary heart disease 8–16, 18
ecogenetics of hypertention 2–4
EDRF 27
endothelin 27
environmental factor 3, 4, 21, 25, 27, 28, 87, 94, 98, 104, 127, 133–136, 139, 147
essential hypertention 20, 27, 38, 39
ethnic differences 2–4, 31, 115–117

familial aggregation studies 1, 2, 8, 11, 12, 98, 99
familial combined hyperlipidemia 9, 12, 15, 17
familial clustering of coronary heart disease 98
familial dyslipidemic hypertention 8, 9, 12–15, 18
familial hypercholesterolemia 9, 11, 16, 98, 110, 111, 117–122, 124, 134, 146, 152–156
family history questionnaire 10, 11
family-oriented preventive medicine 106
fibrin 102, 103, 148
fibrinogen 102, 103, 148
fine mapping 56, 57
folic acid 15
fumarate hydratase gene 45, 51, 84

genealogical data (cf also pedigree) 8, 10, 16, 17, 35
gene-environment interactions 4, 94, 103, 104
gene-gene interaction 104, 105
gene mapping in experimental hypertention 38–59
gene mapping in human hypertention 39
gene therapy 152–156
genetic epidemiology of ischemic heart disease and hypertention 127–142
genetic heterogeneity 1, 4, 5
genetic linkage 4–6, 9, 10, 14, 39, 45, 49, 50, 53, 55, 101, 121, 138, 146
genetic models of hypertention 4, 5, 81–97
genetic variability of SHR and WKY 83–85

hereditary moderate hypercholesterolemia 119–122
high density lipoprotein (HDL) 9, 11–13, 15, 101, 110, 117, 122, 123, 129–131, 133, 134, 137
highly heritable cation traits and hypertention 14
Hin dIII 23, 29
HLA 30, 47, 48
homocysteine 9, 15
human genome project 17
hypertention 1–97, 127–142

identification of polymorphic markers for gene mapping in SHR 49–54
immunocytochemistry 64, 65, 67
increased membrane permeability for ions in hypertention 21
in situ hybridization 61, 62, 64, 67
insulin 8, 9, 12, 13
ischemic heart disease 127–142

Japanese population (and coronary heart disease) 110–126

kallikrein 9, 14, 32
kallikrein gene 32, 46, 47, 52

lipid 8, 10, 12–14, 100, 103, 119, 133–137, 146
lipoprotein lipase 8, 13
low density lipoprotein (LDL) 11–13, 15, 16, 98, 100–102, 105, 110, 111, 113, 124, 143, 144, 147, 153
low density lipoprotein receptor (LDLR) gene 98, 100, 101, 105, 110, 111, 117, 118, 120–122, 146, 154
Lp(a) 9, 13, 15, 17, 98, 100, 102, 103, 110–114, 143–151

manic depressive psychosis 5
MED PED (Tracing Medical Pedigrees to Foster More Early Diagnoses and Prevent Early Deaths) 9, 16, 17
"membranopathy" 21, 22
molecular genetics and familial hypertention 27–37
multifactorial inheritance 1, 2, 4

Na^+-H^+ exchange 21, 23
Na^+-K^+ ATP ase pump 9, 14
Na^+-K^+ cotransport 21, 24, 25
Na^+-Li^+ countertransport 1, 4–6, 9, 14, 21, 32
Na^+-Na^+ exchange 21

obesity 8, 13, 14

pathogenetic basis of hypertention 21, 22
PCR 30, 46, 50, 110, 113, 114, 117
pedigree 12, 16, 119–121, 123
phosphoinositol 23
plasminogen 15, 102, 111, 113, 144, 145, 147–149
plasminogen M5 110–114, 124
polycistic kidney 1
polygenic heritability 10, 14
prevention of cardiovascular diseases 93–95, 106, 124
protein kinase C 23
proto-oncogene 23, 25

quantitative trait loci 39, 41–49, 56, 57

renin-angiotensin system 20, 27, 42, 74, 75, 79
renin-angiotensin system genes 33, 34
renin genes 28–31, 41–45, 52, 55, 60–80
renin gene polymorphism 44, 45
restriction fragment length polymorphism (RFLP) 23, 24–31, 41–47, 55, 100, 101, 117, 118, 120–122, 146
reverse genetics approach 38, 41, 48, 49
risk factors (with respect to coronary heart disease) 1, 98–106, 110–125, 129, 138, 143

salt intake and hypertention 2, 3, 28
schizophrenia 5
segregation analysis 10, 14, 30
sex-linked effects on blood pressure 56
single strand conformation polymorphism (SSCP) 50
spontaneously hypertensive mice 48

spontaneously hypertensive rat (SHR) 21, 23–25, 30, 31, 38–57, 81–95
stress 21, 28
stroke 10, 81–97, 131, 132, 135, 137, 138
stroke-prone rat (SHRSP) 81–95

Tag gene 63, 65–67, 70, 71
TaqI 29, 117
transgenic mice 45, 60–73
transgenic rat 45, 74–80
twin 2, 5, 12, 13, 27, 99, 100, 104
type III hyperlipidemia 9, 15, 115, 124

underdeveloped populations and hypertension 3
uric acid 9
USSR (and genetic epidemiology) 127–142
Utah (control cardiovascular disease in) 8–19, 29, 32

variable number of tandem repeats (VNTR) 35, 44, 45, 50–53

Watanabe heritable hyperlipidemic rabbit 154
Wistar-Kyoto normotensive rat (WKY) 23–25, 31, 40–44, 49, 56, 82–91
whole genome approach 34, 35
World Health Organization 16–18, 94, 128